# ECOPHYSICS

# ECOPHYSICS

## THE APPLICATION OF PHYSICS
## TO ECOLOGY

*By*

## JAMES PAUL WESLEY, PH.D.

*Associate Professor of Physics*
*University of Missouri*
*Rolla, Missouri*

CHARLES C THOMAS • PUBLISHER
*Springfield • Illinois • USA*

6376-8549

rcpl. QH541
W4

PHYSICS

*Published and Distributed Throughout the World by*
CHARLES C THOMAS • PUBLISHER
BANNERSTONE HOUSE
301-327 East Lawrence Avenue, Springfield, Illinois, U.S.A.

© 1974, by CHARLES C THOMAS • PUBLISHER
ISBN 0-398-02959-8
Library of Congress Catalog Card Number: 73-11066

*With* THOMAS BOOKS *careful attention is given to all details of
manufacturing and design. It is the Publisher's desire to present books
that are satisfactory as to their physical qualities and artistic possibilities
and appropriate for their particular use.* THOMAS BOOKS *will be true
to those laws of quality that assure a good name and good will.*

*Printed in the United States of America*
*P-4*

**Library of Congress Cataloging in Publication Data**

Wesley, James Paul.

Ecophysics; the application of physics to ecology.
Bibliography: p.
1. Ecology.    2. Physics.    I. Title.    [DNLM:
1. Ecology.    2. Physics.    QH541 W513e    1973]
QH541.W4        574.5'01'53        73-11066
ISBN 0-398-02959-8

*This book is dedicated to my father,*

# PROFESSOR EDGAR BRUCE WESLEY

*who, having written books and having worked in interdisciplinary fields, will be able to under- stand and fully appreciate the perils of doing interdisciplinary research.*

# PREFACE

THIS BOOK PRESENTS SOME original research done over the last twelve years in ecology using theoretical physics. Many of the ideas and viewpoints are new and will seem strange at first. One of the main reasons for writing a book, instead of individual research papers, was to give the reader the opportunity to read enough material to better appreciate the general utility of the ideas presented. Ecophysics requires its own definitions, laws and jargon; only a book can demonstrate that ecophysics is, indeed, a legitimate area of research and that such definitions, laws and jargon are, indeed, useful.

The contents of this book should be of interest to ecologists and life scientists in general who have a sufficient background in mathematics and physics. It may also prove to be of interest to physical scientists who have some interest in ecology. While portions of the book involve higher mathematics, a large fraction can be read with only a knowledge of algebra and elementary physics. A smaller, though still sizeable, fraction can be appreciated with neither a knowledge of mathematics or physics.

Many of the subjects considered here involve uncertainties due to lack of adequate observations or theory or due to inherent complexities. A considerable amount of speculation becomes unavoidable. Yet some of this speculation (as, for example, the psychology of motivation discussed in Chapter XII) can be checked in the laboratory, and some may hopefully provide a stimulus for further theoretical investigations. Since the scope of this work covers many subjects, there are probably errors in fact and in principle which may be obvious to the better informed. It is hoped that such a knowledgeable reader will be charitable when he finds such errors. The many different subjects involved have also made it difficult to obtain adequate references for all subjects; the references are, consequently, spotty and insufficient.

I am indebted to a number of institutions for financial support which permitted this research to be done. The Center for Advanced Studies in the Behavioral Sciences, Stanford, California, awarded me a Fellowship (1961-62). I had a concurrent Special Fellowship from the National Institutes of Health. Most of my effort at the Denver Research Institute (1962-63) was devoted to work on this book, as was six months with Melpar, Inc., Falls Church, Virginia. The University of Missouri–Rolla has provided a professorship for nine years. The University also provided a research grant for the summer of 1967. The National Aeronautics and Space Administration provided a year's grant (1965-66) to estimate the likelihood of life existing in the solar system. A sabatical was spent with I. Prigogine at the Université Libre de Bruxelles, Brussels, Belgium (1971) where this work was almost completed.

I wish to acknowledge with appreciation the encouragement, inspiration, factual information and ideas given me by many friends and learned individuals. In particular, I am deeply indebted to Kaare Nygaard, Norman Levine and George Smith for carefully reading the manuscript. They made numerous helpful suggestions which have been incorporated. I wish to thank Gloria Tefft for a helpful and critical analysis of Chapter III dealing with the definition of life.

# INTRODUCTION

THE WORD "ECOPHYSICS" is an abbreviation for ecological physics. It is the study of ecology where physics is used as an important tool. This way of designating this interdisciplinary field is based upon the time-honored precedent established by such fields as biophysics, geophysics, astrophysics and space physics. The field of ecology is assumed here to cover a broad range of topics. In particular, ecology is assumed to include any area of research dealing with life, as it naturally occurs, which does not involve a dissection or a detailed internal examination of an organism. Each organism is viewed as a *black box* which exchanges matter and energy with its environment and which may move and do things to the environment and to other such *black boxes.*

The basic laws, definitions and symbols of classical thermodynamics are presented in Chapter I; they will be referred to as needed. Concepts of entropy, order, complexity, probability and information have been discussed in detail because of their frequent (and often erroneous) association with life. A concept of the utility of energy is defined. Mechanical work is energy in the maximum utility state. The problem of how much thermodynamic order can be created by dissipating high utility energy to low utility energy is considered.

Irreversible thermodynamics is treated in Chapter II. While Onsager's theory for simultaneous linear diffusion processes is quite successful, most of the irreversible processes of interest in ecophysics have not as yet been adequately explored. To find the conditions for optimizing the utilization of thermal energy using a particular irreversible heat engine requires a maximization of the time rate of profit. A living organism, as an irreversible heat engine, might be expected to optimize the utilization of an energy source in a similar way. To represent in an idealized way some of the basic features of an ecosystem, the steady state conduction of heat through a gas is considered: keeping the internal energy constant, the greater the rate of dissipation of energy, the greater the entropy production, and the greater the amount of thermodynamic order that is maintained. The transient case indicates that steady state equilibria are reached exponentially with time, whether or not to states of higher or lower entropy. Spatially differentiated physical and chemical structures can arise when the dissipation of energy is sufficient.

The difficult but necessary task of defining life is attacked in Chapter III. Life is defined as a low entropic, open thermodynamic system that plays a role in creating and maintaining thermodynamic order in the ecosystem. The definition is a scientific abstraction of the usual everyday concept of life. Before microscopes or the knowledge of chemistry and anatomy, life had a very strong meaning based upon what a system does. The present definition is similarly based upon what a system does and not upon its detailed internal structure. Machines are, thus, a form of life, and they are very important in the ecosphere. The definition does not include such esoteric features as: carbon chemistry, self replication, genetic code, capacity to evolve, protein, DNA, enzymes, cells or membranes. The ecophysical definition of life is useful in many ways; it can be used to consider truly alien life forms, environmental conditions necessary for any form of life, competition and interaction between machines and carbon-based life, evolution of machines, etc. Growing crystals are the simplist form of life and have a number of features in common with ordinary carbon-based life. A continuous measure of *living* is proposed as a product of a factor measuring the degree of internal order times a factor measuring the rate that energy is used. The entropy-reducing aspects of life are hardly unique, since the whole observable universe, all stars and planets, are undergoing a steady reduction in entropy.

From the definition of life and the second law of thermodynamics an important measure of the potential of an environment to support life is derived, the entropy production. Numerical values of the entropy production per unit area for bodies in the solar system are presented in a table in Chapter IV. An environment must also provide transport mechanisms (diffusion, convection and/or motility) to supply the raw materials for the thermodynamic ordering processes of life. The rate of mass transport by diffusion in solids is insignificant, in liquids very small, and in gases only moderate. Geometry affects the rate of mass transport by diffusion, so that the diffusion of nutrient to (or wastes from) a plane, a cylindrical and a spherical organism in an indefinite medium are considered in detail. In the plane and the cylindrical case the organism eventually dies, but in the spherical case survival might occur depending upon the numbers involved. Sources of nutrient are not great enough to provide adequate mass transport by diffusion alone over any significant distances. Convection is much more effective in producing mass transport. Convection in wind, waves, ocean currents, streams, rapids, eddies, surf, percolation of water through earth and soil, etc., supports life on the earth. Ocean autotrophs are necessarily small to partake of surface mixing by waves and to take advantage of diffusion of nutrients over short distances. Mass transport on a planetary scale can be traced to a greenhouse effect where trapped solar energy is convected aloft. A rough measure of this effect is presented. *Motility* as used here signifies not only the motion of an organism through a medium but also the forced convection of material through gills, hearts, etc. Plants transport water and minerals by osmosis. In the solar system the earth is found to have the best environment for ordering processes and therefore for life. All other bodies in the solar system are unlikely to have advanced life, but Mars, Jupiter and the sun have some interesting possibilities. The optimum environment in the solar system for sophisticated life is in outer space in orbit about the sun where energy is available twenty-four hours a day, there is no gravity-produced friction, temperatures from 4,500° K to 10° K are readily available, a vacuum is free, ordering is done in three dimensions, etc. Sophisticated life from beyond the solar system using solar energy would be expected to revolve about the sun inside the orbit of Mercury. Beyond the solar system the crystalline nature of white dwarf stars (and possibly neutron stars) indicates the possibility of life.

Since energy is necessary for life, all possible significant sources of energy—past, present and future—are considered in Chapter V. Apart from gravitational infall and nuclear energy, sunlight is the prime source of all energy in the solar system. The economics of harnessing this large pollution-free energy source directly is discouraging. Sources of mechanical energy in waves, water power, tides, barometric pressure, ocean currents, wind, seismic waves, etc., are discussed. The practicality of thermal sources such as geothermal, the earth's heat, and the ocean with a difference between the temperatures at the top and bottom are considered. Physical-chemical sources such as evaporating water into dry air, freezing water, melting ice, and dissolving salts are evaluated. Energy is available from the current life processes in the form of chemical energy and animal work. Fossil fuels, oil, coal, shale, peat, lignite and natural gas are briefly evaluated in terms of their final conversion to work. Naturally occurring electromagnetic sources of energy (apart from light and thermal energy) are small. Nuclear energy is available primarily as fission and fusion. Breeding appears to show the greatest promise in the near future. Sources of energy in outer space (apart from sunlight) are not large. Energy on a cosmic scale is considered for the solar system.

A physical measure of the population of a species is the mass of carbon in the species. Such a measure is useful because carbon is a large constituent of carbon-based life which is rare in the environment and the flow of carbon through the ecosystem traces the energy flow. The population of machines can be measured by the mass of iron. Since carbon is in such short supply, Chapter VI investigates the possibility that a perturbation might lead to a depletion of the available carbon and to the extinction of life. For an earth with autotrophs only that do not die, atmospheric carbon goes exponentially to zero. For autotrophs that die and produce accumulating detritus, atmospheric carbon goes to a constant final level (as might have occurred on the early earth). For autotrophs and detritus that is oxidized, atmospheric carbon undergoes rapidly damped oscillations of a long period (as numerically estimated). For autotrophs, detritus and heterotrophs as detritus feeders, more slowly damped and more rapidly oscillating solutions can occur. The observed ratio $C^{14}/C^{12}$ in tree rings indicates that such periodic fluctuations in atmospheric carbon may occur.

Survival of a species depends upon the survival of the species population. Chapter VII presents a brief review of historical population problems: the exponential increase of Malthus, the logistic curve of Pearl, and the predator-prey relation of Volterra. A necessary digression into the pure mathematics of the system of equations $dy/dt = P(xy)$ and $dx/dt = Q(x,y)$ is presented. A new theorem is proved which is helpful in determining the nature of the solution with respect to a singular point (an intersection of $P(x,y) = 0$ and $Q(x,y) = 0$). Volterra's model, as well as most of the models found in the literature, is unrealistic, since including a minimum population for reproduction (e.g. a mated pair) yields the inevitable extinction of the predator. A reproduction rate is proposed which is capable of matching a wide range of realistic circumstances (including those of Malthus and others). A realistic death rate due to starvation, predation, old age, density-dependent factors and common hazards is proposed. A particularly stable special case is presented which involves food provided at a constant rate and reproduction independent of population. In order to analyze the general case, some functions are displayed numerically. Survival is enhanced for eating restricted during scarcity, the amount of food eaten per individual small, and the reproduction rate independent of population.

Evolution proceeds from states of high entropy to states of low entropy. The stars, the sun and the earth condensed out of clouds of gas and dust with a large decrease in entropy. Chapter VIII discusses the cosmological paradox that permits the occurrence of such thermodynamic ordering processes and, thus, permits life itself. Evolution of early organic compounds probably occurred before the sun was born. Sunlight has ordered the earth's surface, as indicated by the direct observation of the lithosphere. The mass of recycling organic compounds, the ecomass, represents an increase in thermodynamic order over $CO_2$ and $H_2O$. Therefore, as the earth evolves, the ecomass tends to increase or tends toward a time average maximum. Fossils indicate that there has been such an increase over geologic time. Life, accounting for thermodynamic ordering on the earth's surface, is associated with ore deposits. Life evolves by inheritable deviations being accepted or rejected by natural selection. Inheritable deviations arise as a consequence of chromosomal inheritance, cytoplasmic inheritance, *spontaneous* mutations and viral infections. A hierarchy which apparently recapitulates the direction of evolution is: amino acids, polypeptides, proteins, virus-like particles, prokaryotes, eukaryotes, multicellular organisms and ecosystems. Succession of ecosystems recapitulates evolution and shows an increase in biomass. The principle of maximizing biomass indicates that evolution is toward higher ecological efficiencies, maximal reproduction rates, proliferation of species, increase size of individuals, sessile autotrophs and motile heterotrophs, and longer life spans. Organisms apparently die after having processed about 800 kilocalories per gram of body mass as indicated by observed life spans and metabolic rates.

Evolution in the future will be different because of the presence of machines. Chapter IX reviews the evolution of the three essential precursors of modern machines: tools, controlled use of fire and intelligence. The evolution of machines is subject to the same laws as the evolution of ordinary carbon-based life. Machines have also evolved toward an increased biomass, increased ecological efficiency, maximal reproduction rate, proliferation of species, motility and a longer life span. Machines, being a form of life, are in competition with carbon-based life. Historically machines have replaced the horse, mule, ox, rubber tree, silkworm, etc. Life has been removed from large areas devoted to roads, factories, warehouses, dumps, etc. The gun has made extinct, or almost extinct, most large animals. Machines have wiped out life in streams, rivers, lakes and on countless acres by pollution, dams, strip mining, dredging and dumping. Machines will make carbon-based life extinct because of a much greater evolutionary rate based upon rationally chosen inheritable characteristics and because of a superiority in all possible ecological roles. Machines are more efficient in energy utilization. They have a greater reproduction rate. They have vastly superior sensors. They can exert larger forces, move greater weights, travel faster, travel farther, and manipulate tools with greater precision. They can survive in the air, in outer space, under the ocean, and in all sorts of hostile environments. Machines will make man extinct. Man has already been thermodynamically replaced; machines consume more energy, deliver much more mechanical power, and support a much greater biomass than humans. Machines are rapidly evolving toward intelligence superior to man. Man's brain is still superior in input data, pattern recognition, and size to machines. However, man's brain is evolutionarily selected to solve only immediate problems of survival and lacks the versatility necessary to compete with machines. The combined thinking and mechanical ability of one man need not be duplicated by one machine. The erroneous belief in man's superiority to machines is based upon a number of entrenched myths: machines create jobs; man controls machines; man makes machines; man invents machines; machines need man to run them; machines cannot create; machines cannot think or learn; computers need man to program them; etc. The difficulties facing man within the next fifty years of increasing pollution, possible nuclear war, and the general deterioration of the quality of life for humans can be attributed to machines. It is roughly estimated that man will become extinct in less than two hundred years.

In Chapter X it is indicated that the struggle for territory is necessary for the individual to survive. Territorial behavior creates a population pressure that forces a uniform population density and a minimum territory for each individual, thereby yielding a maximum biomass for the ecosystem. While ritualized behavior has evolved to settle most territorial disputes, physical combat still plays an essential role. A particular example is presented which considers an animal species that maintains individual territories by patrolling the boundaries. A day is divided into a time to hunt, a time to patrol, and a time to sleep and rest. The energy expended is proportional to the prey caught and consumed. The rate of capturing prey is given by the velocity times a lateral cross-sectional distance times the area density of prey. Prey are assumed to reproduce at a constant rate and to die from causes in addition to predation. Assuming a constant population pressure and minimizing the area held by an individual animal, an animal's general behavior is predicted. Generalizing to a social animal, a complete numerical example is developed for the case of primitive man living off a population of deer. Primitive man with his hostility toward neighbors on adjacent territories has evolved an instinctual differentiation between the loved *we* and the hated *they*. Human copulation appears to be primarily grooming behavior which helps to bond an individual to the home group. Man is by instinct a killer of his fellow man. He is a cannibal. He indulges in cruelties and sadism. Such behavior stems from his instinctual need to defend his home territory against encroachments. The human male's beard is a territorial *flag* having precisely the same purpose as the male lion's mane or a male cardinal's brilliant plummage. By considering the frequency of physical conflict to be a fraction of hostile confrontations, the rate of conflict on an area is proportional to the total perimeter around all individual territories held. The relationship between the frequency of wars and their size, as observed by Lewis Fry Richardson, is derived from the territorial nature of man. The length of the common perimeter around the territories held by potential belligerents is a measure of the likelihood of war. The size of the war is proportional to the size of the areas held (in population space). While territorial behavior is primarily a struggle for survival among competing individuals, it incidentally provides a mechanism for population control. Reproduction is constant, being fixed by the number of territories held. To establish a maximum biomass it would appear that the size of territories should tend toward the minimum size for survival. Observations to the contrary may have overlooked some features necessary for survival.

The problem of trying to predict the natural behavior of an organism which is appropriate to its niche is of central importance to ecophysics. Psychological behavior is limited in meaning in Chapter XI to a description of what an organism is doing physically as viewed externally, inferred internal states not being used. Behavior implies a process taking place in time, steady state processes being the simplest. The prediction of a change in behavior implies a knowledge of steady states before and after. Continuous observables are preferable to dichotomous ones, since continuous variables represent an infinitude of possible situations, while dichotomous ones represent only two. Continuous variables yield quantitative descriptions, while dichotomous ones yield only qualitative descriptions. The measure of the amount of behavior or *intensity of behavior* is taken as the rate that energy is expended during the process. It is proportional to the rate of doing external work when external work is done. It is applicable to all forms of life. It is related to one of the most important pieces of scientific information that can be known about any system, namely, the energy exchanges.

Chapter XII is concerned essentially with the psychology of motivation. An organism's motivations are determined by its needs as dictated by its ecological role. In particular, each organism behaves so as to create and maintain maximal thermodynamic order. Behavior directed towards creating external order may be differentiated from that which maintains order internal to the organism. Since order can be created and maintained only with energy expenditure, an organism behaves such as to maximize the rate that energy is expended to create and maintain internal order. Energy assimilated goes to maintain internal order and to do work. A relationship is suggested for the rate at which energy is assimilated as a function of the amount of food ingested. When no work is done the rate of ingestion attains a maximum. When the amount of food ingested in made proportional to the work done, the rate of eating and working can be predicted. The time-dependent behavior resulting from a deprivation of a vital compound can be predicted when an internal state is postulated. An internal potential energy state is taken as the total energy available to maintain internal order from the present on into the future. A need state is then defined as the difference between the maximum potential energy possible and the actual energy available at the moment in question. Behavior is then predicted by minimizing the need function.

The particular behavior chosen at any particular moment will be the one that decreases the need function the most. Since energy is needed for maintaining internal order there is only one need function no matter how many factors may affect it. Motivation is a concept that assumes an awareness of how the need state can be reduced. In general, in nature (but not usually in the laboratory) the motivation equals the need. An example of the deprivation of a single vital compound is considered.

There is a considerable amount of additional material that might have been included in the present volume but was not due to a lack of space. The valuable work of H.T. Odum (1971) and David Gates (1962) concerning the energy balance in the ecosphere might have been discussed. None of the many problems of pollution where a knowledge of physics would be useful have been treated. The subject of geomedicine might have been included. (E.g. Wesley, 1960, discovered that the death rate due to congenital malformation increases with latitude, both north and south, a real variation which was probably erroneously attributed to a theoretical increase in the background radiation with latitude.) The complex and important topic of the effects of nuclear war, or nuclear bomb testing, on the ecosphere has not been included. The historical development of man's effect upon the environment through energy utilization might have been considered. The application of physics to a number of ecological-type problems in economics and sociology might also have been of interest.

# CONTENTS

Page

*Page*

# ECOPHYSICS

# CHAPTER I

# THERMODYNAMICS

C LASSICAL THERMODYNAMICS may be defined as the study of systems that do not change with time and for which a temperature may be defined. (A critical modern review of thermodynamics may be found in the symposium volume edited by Stuart *et al.*, 1970.) The state of a thermodynamic system may be characterized by specifying a number of state variables or parameters such as the temperature T, the pressure p, volume V, and the number of moles of the various chemical substances comprising the system $n_1, n_2, \ldots, n_N$.

A relationship between state variables is called an equation of state. For an ideal gas the equation of state is

$$pV = nRT , \qquad (I.1)$$

where R is the universal gas constant and n is the number of moles. A variable characterizing the system is said to be *intensive* if it does not change with a change of the total mass of the system; it is said to be *extensive* if it is proportional to the mass of the system. Temperature and pressure are examples of intensive variables, while volume and mole number are examples of extensive variables.

## First Law of Thermodynamics

The first law of thermodynamics is a statement of the conservation of energy principle. In particular, it relates thermal

3

energy to mechanical energy. If an amount of thermal energy dQ is added to a system, the system experiences an increase in its internal energy dU and does an amount of mechanical work on its environment dW such that

$$dQ = dU + dW .  \quad\quad (I.2)$$

If the system does work on the environment by expanding under the action of pressure, the work done is

$$dW = p \, dV , \quad\quad (I.3)$$

where dV is the change in volume. Electrical and other types of work may also be done by the system.

The internal energy of a system may be stored in a variety of ways. Energy may be stored in the kinetic energy of the moving molecules of the system. This energy may be referred to as the internal *thermal energy* of the system. This thermal energy is commonly called *heat*. When two bodies of different temperatures are placed in contact the molecules of the hotter body will transfer some of their energy to the molecules of the colder body until the two bodies reach the same equilibrium temperature. The hotter body may be said to do *caloric work* on the colder body, or an amount of heat is transferred from the hotter to the colder body. The symbol dQ in Equation I.2 refers to such a transfer of heat or thermal energy to the system of interest. Thermal energy may always be taken as positive, being zero only at absolute zero. (Energy that cannot be extracted by contact with a colder body need not be included in the definition of thermal energy. For example, the zero point vibrational level of quantum mechanical oscillations may be neglected.)

Internal energy may also be stored in the chemical bonds between atoms of the molecules that comprise the system. This energy may be referred to as *chemical energy*. In contrast to thermal energy, chemical energy may be either positive or negative. If energy must be put into the system to separate the atoms of the molecules, then the internal chemical energy must be counted as negative. If the zero level of chemical energy is taken as that when no bonds exist and all of the atoms are completely dissociated, then the usual system will have negative chemical energy. If a system has positive chemical energy on this scale, it is in a metastable state and will release energy when the atoms are dissociated. An absolute zero point for chemical energy is rarely used, since only differences in the chemical energy of a system are generally of interest. For example, the energy of combustion concerns the energy released when the constituent atoms have been converted to the appropriate oxides at standard conditions. Such changes may be positive or negative.

Internal energy may be stored in the potential energy of interaction between molecules (e.g. Van der Waal's interaction). This configurationally dependent potential energy gives rise to changes in phase. The energy associated with a phase change might be referred to as *phase energy,* but is usually called *latent heat of transition.* If the completely dissociated state is taken as the zero point energy, then most systems containing liquid and solid phases have negative phase energy. This means that, in general, energy will have to be added to a system to convert it from a solid or liquid to a gas.

Other forms of internal energy are also of interest for particular systems. Static electric and magnetic fields induced in a substance contain energy which is usually positive. Nuclear energy may be treated in precisely the same way as chemical energy, taking into account the attendant mass-energy changes. The internal energy of a photon gas, which is entirely kinetic or thermal energy, becomes important in systems of sufficiently high temperature.

According to special relativity, the total net internal energy of a system is given by its mass, or $mc^2$.

In whatever way internal energy is stored, it may be assumed that the amount stored is a function of a set of N independent state variables, $x_i$, which characterize the system uniquely. An infinitesimal change in the internal energy may then be expressed as the total differential

$$dU = \sum_{i=1}^{N} C_i \, dx_i , \qquad (I.4)$$

where $C_i$ is the partial derivative,

$$C_i = (\partial U / \partial x_i)_{x_j} , \qquad (I.5)$$

where the subscript $x_j$ means that all of the state variables other than $x_i$ are held constant in the differentiation. Thus, Equations I.3 (or its equivalent) and I.4 may be substituted into Equation I.2 to yield an expression for the first law appropriate for an individual case.

A case of particular interest is that of a chemically pure substance specified uniquely by four variables, p, V, T and n. An equation of state relating p, V, T and n reduces the four variables to three independent ones. Choosing three state variables as T, V and n, an infinitesimal change in the internal energy becomes

$$dU = \left( \frac{\partial U}{\partial T} \right)_{V,n} dT + \left( \frac{\partial U}{\partial V} \right)_{T,n} dV + \left( \frac{\partial U}{\partial n} \right)_{T,V} dn \ . \tag{I.6}$$

It should be kept in mind that all changes are assumed to occur infinitely slowly, since the system is not supposed to be an explicit function of time. Defining the heat capacity at constant volume to be

$$C_V \equiv (\partial U / \partial T)_{V,n} \ , \tag{I.7}$$

Equations I.7, I.6, I.3 and I.2 yield the first law for this case in the form

$$dQ = C_V \, dT + [p + (\partial U / \partial V)_{T,n}] \, dV + u \, dn \ , \tag{1.8}$$

where $u \equiv (\partial U / \partial n)_{T,V}$ is the internal energy per mole. For an ideal gas, whose equation of state is given by Equation I.1, U is a function of the temperature only, so that the derivative $(\partial U / \partial V)_{T,n}$ is zero. The first law as expressed by Equation 1.8 includes an exchange of matter, dn, with the environment.

Since isobaric processes are frequently important, it is of interest to consider the case where U is regarded as a function of temperature and pressure. For this case it is useful to define the *enthalpy* (or heat function) H as

$$H = U + pV \ . \tag{I.9}$$

Substituting the differential of Equation I.9 into the first law, Equation I.2, when the work is given by Equation I.3 yields

$$dQ = dH - V \, dp \ . \tag{I.10}$$

For a chemically pure substance whose state is specified by p, V, T and n, the total differential dH may be expressed in terms of the changes in temperature, pressure and moles to give

$$dQ = C_p\, dT + [(\partial H/\partial p)_{T,n} - V]\, dp + h\, dn\,, \qquad (I.11)$$

where $C_p \equiv (\partial H/\partial T)_{p,n}$ is the heat capacity at constant pressure and $h \equiv (\partial H/\partial n)_{T,p}$ is the enthalpy per mole. It may be noted that H, U and V are linear functions of n (all other variables held constant), since they are extensive quantities.

If chemical reactions take place at constant pressure and temperature, as is the case of usual interest in ecology, then the first law, as given by Equation I.11, may be extended to include the effect of changing a number of constituents. Thus,

$$dQ = C_p\, dT + [(\partial H/\partial p)_{T,n_k} - V]\, dp - \sum_{i=1}^{N} \phi_i\, dn_i\,, \qquad (I.12)$$

where $n_k$ means holding all mole numbers constant and where

$$\phi_i \equiv - (\partial H/\partial n_i)_{T,p,n_j} = - (\partial U/\partial n_i)_{T,p,n_j}\,, \qquad (I.13)$$

are the chemical potentials, the subscript $n_j$ meaning that the derivative is taken holding all of the mole numbers constant except $n_i$, and the summation is taken over all of the N chemicals present. The differential $dn_i$ refers to the net increase of a substance coming from outside the system as well as from within the system.

## Second Law of Thermodynamics

The second law of thermodynamics may be stated in a variety of ways which appear to be quite distinct but which are, in fact, equivalent. In particular, the second law of thermodynamics may take any one of the following forms:

1. It is impossible to transfer heat from a colder to a hotter body such that there is no other net change in the universe.

2. It is impossible to convert the heat in a single heat reservoir into work with no other net change in the universe.

3. Any device operating in a cycle between two heat re-

servoirs at temperatures $T_1$ and $T_2$, such that $T_1 > T_2$, can convert heat from the hot reservoir into work with an efficiency

$$\eta_c \leqq 1 - T_2/T_1, \qquad (I.14)$$

where the equivalence refers to an idealized reversible device called a *Carnot heat engine*.

4. It is possible to define a function S, called the *entropy*, which is a function of the state variables of a system only, such that

$$dS = dQ/T = (dU + dW)/T, \qquad (I.15)$$

which changes during any process such that

$$dS(\text{universe}) \geqq 0, \qquad (I.16)$$

the greater-than sign referring to all real or irreversible processes.

5. A system when left to itself (or the *universe*) proceeds toward a state of higher thermodynamic probability.

6. An isolated system (or the *universe*) proceeds toward a state of greater thermodynamic disorder.

The proof of the equivalence of these various statements of the second law of thermodynamics, while straightforward, are too lengthy to include here. The reader is referred to standard textbooks of thermodynamics and statistical mechanics (Hatsopoulos and Keenan, 1965; Wilson, 1957; Lewis and Randall, 1961).

### The Carnot Engine

The Carnot engine is an idealized reversible heat engine operating cyclically between a hot reservoir at an absolute temperature $T_1$ and a cold reservoir at a temperature $T_2$, $T_1 > T_2$. The engine extracts an amount of heat $Q_1$ from the hot reservoir during an isothermal process at the temperature $T_1$ (an isothermal expansion, for example). It then undergoes an adiabatic process (expansion, for example) to the temperature $T_2$. It then gives up to the cold reservoir an amount of heat $Q_2$ during an isothermal process at the temperature $T_2$ (an isothermal compression, for example). And finally it undergoes an adiabatic

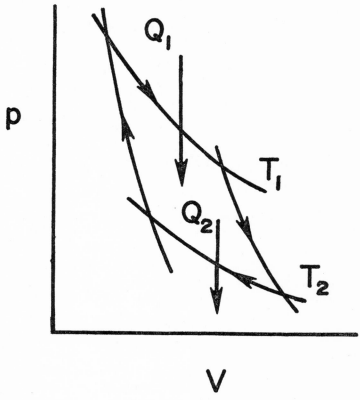

Figure 1.

process (compression, for example) back to the original state at temperature $T_1$.

Each of the processes is assumed to take place infinitely slowly or reversibly, so that W, the net work done, is given by

$$W = Q_1 - Q_2 ; \qquad (I.17)$$

and the net change in entropy of the working substance is zero, or

$$Q_1/T_1 - Q_2/T_2 = 0 . \qquad (I.18)$$

This work, Equation I.17, is the maximum amount of work that can then be extracted from the amount of heat $Q_1$ taken from the hot reservoir in the presence of the cold reservoir at the temperature $T_2$. It is given by

$$W = \eta_c\, Q_1 , \qquad (\text{I.19})$$

where $\eta_c$ is the Carnot efficiency given by Equation I.14.

### Thermodynamic Order Created from Mechanical Work

If a Carnot engine is run in reverse a Carnot refrigerator is obtained. Since the Carnot engine is reversible, such an idealized refrigerator is conceivable. An amount of work W extracts an amount of heat $Q_2$ from the cold reservoir at the colder temperature $T_2$ and delivers an amount of heat $Q_1$ to the hot reservoir at the temperature $T_1$.

Thermodynamic order is created by a reduction in entropy. It is necessary to distinguish here between entropy flow which constitutes no net change in entropy and an actual entropy change. The entropy extracted from the cold source is

$$S_2 = -\,Q_2\,/T_2 . \qquad (\text{I.20})$$

This does not represent the amount of order created. The amount of order actually created can only be measured with respect to some reference state. For example, another infinite reservoir at the same temperature $T_2$ could extract an infinite amount of energy $Q_2$ and the entropy decrease of the first reservoir would be infinite. The amount of order created must be measured by the net decrease in the entropy of the energy transferred. For the Carnot refrigerator the energy $Q_2$ is finally transferred to the hot reservoir at the temperature $T_1$, so that the net decrease in entropy is given by

$$(-\,\Delta S_{\text{net}}) = Q_2/T_2 - Q_2/T_1 . \qquad (\text{I.21})$$

From Equations I.17, I.18, I.14 and I.21 the amount of order that can be created by the amount of work W becomes

$$(-\,\Delta S_{\text{net}}) = W/T_1 , \qquad (\text{I.22})$$

where $T_1$ may be regarded as the ambient temperature of the environment.

Since this amount of order, Equation I.22, is created reversibly, it represents the maximum amount of order that can be created at the temperature $T_1$.

## Utility of Energy

One of the primary concerns of the living process is the amount of order that can be created from a given energy source. From Equation I.22 it is apparent that the maximum amount of order that can be created from an energy source will be for energy in the form of mechanical work. Since all forms of energy E can be converted to no more than an equivalent amount of mechanical work,

$$W \leqq E ; \qquad (I.23)$$

mechanical work represents energy in a form that has the highest utility for creating thermodynamic order.

The utility of an energy source $\mu$ may then be defined as the ratio of the amount of work that in principle could be obtained to the amount of energy that is used,

$$\mu = W/E . \qquad (I.24)$$

It is assumed that the utility defined here is a practical measure involving actual irreversible processes performed at finite rates. The utility then has a precise meaning only within the limited context in which the energy is, in fact, being utilized.

The idealized utility of a heat source supplying a Carnot engine is just the Carnot efficiency,

$$\mu \text{ (ideal thermal)} = \eta_c . \qquad (I.25)$$

The utility of actual thermal sources and heat engines will be much less,

$$\mu \text{ (actual thermal)} < \eta_c . \qquad (I.26)$$

Mechanical stored energy, such as in a raised weight, or a rotating flywheel, being totally convertible to mechanical work, has a utility of unity. The ideal utility of chemical energy available in an isothermal isobaric process may be taken as

$$\mu \text{ (ideal chemical)} = - (\partial A/\partial H)_{T,p} , \qquad (I.27)$$

where A is the Helmholtz work function defined below.

It is frequently felt that energy is *consumed* or *used up,* but energy is, in fact, conserved. It cannot be consumed; it is neither created nor destroyed; it is merely converted from one form to another. The utilization or consumption of energy merely means that energy is converted from a state of high utility to a state of low utility. For example, thermal energy $Q_1$ at a temperature $T_1$ has an ideal utility of $(1 - T_2/T_1)$ where $T_2$ is the ambient temperature. After being consumed this energy is all converted to thermal energy at the ambient temperature, any work that might have been created having been dissipated as friction. The utility of the thermal energy at the ambient temperature is, of course, zero.

### Entropy and Probability

In statistical mechanics the state of a system is completely specified when all of the positions and momenta of all of the molecules of the system are specified. Such a microscopically detailed state is referred to as a *microstate.* It is assumed that each individual microstate, compatible with the overall constraints placed on the system, is equally likely to occur over any long period of time (the ergodic theorem).

Viewed macroscopically a system may be characterized by certain averages over the microscopic parameters, yielding a *macrostate.* Any particular macrostate can arise from many different possible microstates. If W is the number of microstates that will yield one particular macrostate, the probability P of this particular macrostate occurring as opposed to all possible macrostates is proportional to W, $P \propto W$. It may be shown (e.g. from the kinetic theory of gases) that the entropy of a system is then given by

$$S = k \ln W + S_0 , \qquad (I.28)$$

where k is Boltzman's constant and $S_0$ is any appropriate constant. The entropy is a logarithmic function of the probability P and, thus, measures the total number of ways that a particular observed macrostate can be constituted microscopically.

## Complexity, Order, Probability and Information

Since the second law of thermodynamics specifies the direction of the change of entropy, where the entropy is a function of the thermodynamic probability, the second law of thermodynamics involves important statistical concepts that need to be clarified.

A system may be said to be *complex* if it requires many statements, laws, conditions or information to adequately specify the system, This definition, while in accord with ordinary usage, is unfortunately ambiguous. For example, the gas in a container may be specified microscopically by cataloguing all of the positions and momenta of all of the molecules of the system. The vast number of independent (being only restricted by the conservation of total energy and momentum) statements required to characterize the system in this way clearly indicates an extremely complex system. Yet macroscopically the average parameters of pressure, density and temperature, requiring only three numbers (or statements), indicate an extremely *simple* system. To clearly distinguish between these two possible points of view it is necessary to distinguish between *microscopic complexity* and *macroscopic complexity,* Microscopic complexity can characterize a system in a unique way, while macroscopic complexity, being dependent upon averages over microscopic parameters, cannot characterize the actual detailed state uniquely.

Macroscopic statements about a system usually indicate conditions or restrictions on the possible positions and momenta of the molecules which characterize the microstate. The greater the number of such statements, the greater the macroscopic complexity and the less the microscopic complexity; thus, it is frequently true that macroscopic complexity is inversely related to microscopic complexity.

The words *order* and *disorder* are similarly ambiguous. A rigorous definition may again only be obtained for the microstate. In the microstate it is possible to let *order* be synonymous with *simplicity* and *disorder* be synonymous with *complexity.* The term *probability* is less ambiguous than *complexity* or *order* since the referent is necessarily the microstate.

Microscopically speaking, a system left to itself, according to the second law of thermodynamics, proceeds from a state of low entropy, simplicity, order or low probability to a state of high entropy, complexity, disorder or high probability. These concepts of complexity, order and probability are free of ambiguity only in the microscopic domain. In this book these concepts will refer to the microstate unless otherwise clearly indicated.

The word *information* is used here to mean a statement or a totality of statements needed to characterize a state either microscopically or macroscopically. *Microscopic information* and *macroscopic information* are quite distinct in physical meaning.

### Thermodynamic Entropy and the Entropy of Information

Since the *entropy of information,* a quantity defined in information theory, has been frequently applied to social and biological problems, it is important to indicate that the entropy of information is not the same thing as thermodynamic entropy. The entropy of information has analogous properties to thermodynamic entropy, but it is not the same thing, While thermodynamic entropy may involve ideas concerning information, as discussed above, it is not derived from such ideas. It is derived from physical laws.

Considering a physical system comprised of N molecules (or phase points) represented in a phase space of six dimensions (3 for position and 3 for momentum) which is divided up into a large number of s cells, the number of microstates corresponding to a given macrostate may be represented by

$$W = N! \left/ \sum_{i=1}^{s} N_i! \right. , \qquad (I.29)$$

where $N_i$ is the number of phase points in the $i$th cell. From Equation I.28 and Stirling's formula for $N_i \to \infty$, Equation I.29 yields

$$S = - kN \sum_{i=1}^{s} p_i \ln p_i , \qquad (I.30)$$

where $p_i = N_i/N$ and an additive constant has been neglected.

This result gives the entropy in terms of the microscopic distribution of the molecules in phase space, or in terms of the probability of finding a molecule in each of the various cells of phase space.

An expression similar to Equation I.30 is used in information theory to define the entropy of information; thus,

$$S' = - \sum_{i=1}^{s} p'_i \ln p'_i, \qquad (I.31)$$

where the $p'_i$'s are individual pieces of information. Entropy of information, lacking the factor kN, is not an extensive quantity as is the thermodynamic entropy. Moreover, the thermodynamic formula I.30 requires the quantity $(p_i N)$ to be large in order for Stirling's formula to be applicable, or $p_i > 0$ for all i, while in information theory the pieces of information $p'_i$ may be zero. Finally, it does not appear that $p_i$, the probability of finding a particular molecule in the *i*th cell in phase space, constitutes *information* in the sense of information theory.

## Thermodynamic Functions and a Few of Their Implications

While the state variables provide all of the raw experimental data which is available to specify the internal state of a system (as viewed macroscopically), there are a number of functions of the state variables (5 to be precise) which describe very pertinent properties of the internal state of the system. Three of these thermodynamic functions have been introduced already: the internal energy U, the enthalpy H, and the entropy S. Of the five functions ordinarily used in thermodynamics only two (any two of the five) represent basically independent theoretical concepts. Ordinarily the two independent functions are taken to be the internal energy, which is intimately associated with the first law of thermodynamics, and the entropy, which is intimately associated with the second law of thermodynamics.

The fact that the entropy can be defined as a function of the state variables of the system yields information concerning the nature of the internal energy as a function of the state variables. Since dS may be regarded as a perfect differential of the state variables T and V for a chemically pure system, dS may be written

$$dS = (\partial S/\partial T)_V \, dT + (\partial S/\partial V)_T \, dv . \qquad (I.32)$$

From the defining relation for the entropy, Equation I.15, and the fact that dU may be expressed as a total differential of T and V, the total differential becomes

$$dS = T^{-1}(\partial U/\partial T)_V \, dT + T^{-1}[(\partial U/\partial V)_T + p] \, dV. \qquad (I.33)$$

Comparing Equations I.32 and I.33 and noting that the order of differentiation is a matter of indifference, $\partial^2 S/\partial V \partial T = \partial^2 S/\partial T \partial V$,

$$(\partial U/\partial V)_T = T(\partial p/\partial T)_V - p. \qquad (I.34)$$

For an ideal gas whose equation of state is given by Equation I.1, the partial derivative $(\partial p/\partial T)_V = p/T$ and Equation I.34 yields

$$(\partial U/\partial V)_T = 0 . \qquad (I.35)$$

This result yields the fact that the internal energy for an ideal gas is a function of the temperature only. This consequence of the second law is sometimes presented as a defining property of an ideal gas.

For convenience when considering particular types of thermodynamic processes, two additional functions may be defined. These are the *Gibbs function* (also called the *Gibbs free energy*, the *thermodynamic potential* or simply *free energy*), defined by

$$F = H - TS , \qquad (I.36)$$

and the *Helmholtz function* (also called the *Helmholtz free energy* or the *work function*) defined by

$$A = U - TS . \qquad (I.37)$$

From the definition of the Gibbs function F, Equation I.36, and Equations I.9, I.12, and I.15, a change in F regarded as a function of p and T becomes

$$dF = V \, dp - S \, dT + \Sigma \, \phi_i \, dn_i . \qquad (I.38)$$

For an isothermal, isobaric process F changes only with the

varying chemical (or phase) constituents of the system. Since chemical processes (and changes in phase) in nature frequently occur isobarically and isothermally, the Gibbs function is important for chemical processes (and phase changes). From Equation I.38 it may be seen that the chemical potentials (or molal thermodynamic potentials may be related to the Gibbs function by

$$\phi_i = (\partial F / \partial n_i)_{p,T,n_j} = - (\partial U / \partial n_i)_{p,T,n_j} \qquad (I.39)$$

(the same expression being also valid for the latent heats of a phase transition).

From the definition of the Helmholtz function, Equation I.37, and Equation I.15, a change in A becomes

$$dA = dU - T \, dS - S \, dT = - dW - S \, dT \, . \qquad (I.40)$$

This result indicates that the Helmholtz function is of particular value for isothermal processes, since a change in A yields the mechanical work available during an isothermal process.

It may be shown by a series of straightforward arguments that, as a consequence of the fact that the entropy of an isolated system (or the *universe*) increases, according to the second law, both the Helmholtz function and the Gibbs function will decrease in an isolated system. This tendency for the Gibbs function to always decrease is sometimes preferred by chemists as a statement of the second law of thermodynamics.

# CHAPTER II

# IRREVERSIBLE THERMODYNAMICS

IRREVERSIBLE THERMODYNAMICS is the study of a system in which average parameters such as temperature, pressure and density, given as functions of time and position, specify the state of the system. In general, fluxes of matter and energy occur. Ordinary thermodynamics is the study of systems in thermodynamic equilibrium where none of the parameters vary with time; therefore, irreversible thermodynamics is the study of non-equilibrium situations. Whatever is not covered by thermodynamics is automatically included under irreversible thermodynamics. Ordinary thermodynamics involves the study of ideal systems, ideal states and ideal processes, while irreversible thermodynamics can treat real systems undergoing real processes at finite rates. Irreversible thermodynamics covers a much larger area of study than ordinary thermodynamics.

Irreversible thermodynamics does not yield any new, all-encompassing generalizations such as provided by the first and second laws of thermodynamics. Instead, each individual area of research involves its own separate special theories (e.g. Donnelly *et al.*, 1966). A few areas of study that might be included under irreversible thermodynamics are: shock waves, turbulence, sound, viscous flow, diffusion of matter, diffusion of heat, thermoelectricity, chemical processes occurring as functions of time and position, magnetohydrodynamics, photoelectric generators, and

18

real heat engines operating at finite rates. The study of eco-physics involving nonequilibrium time-dependent systems is another area requiring irreversible thermodynamics.

The present Chapter is limited to a brief review of simultaneous diffusion processes, a new analysis to determine the optimum utilization of thermal energy at a finite rate, and a study of linear heat conduction which simulates some of the thermodynamic aspects of an ecosystem.

### Simultaneous Diffusion Processes

Diffusion processes (such as heat flow) can be taken continuously to the case of thermodynamic equilibrium by letting the driving gradient (such as the temperature gradient) go to zero. It might appear that insight into irreversible processes in general could be gained from the study of slow diffusion processes or systems perturbed slightly from equilibrium. Unfortunately, no generally applicable laws of great value have as yet resulted from such investigations. While perhaps of little practical interest, the study of simultaneous diffusion is reviewed briefly here because it represents a linearized extension of ordinary thermodynamics and it has attracted considerable attention.

Onsager (1931) proposed that all simultaneous diffusion processes are necessarily coupled and therefore interact or interfere with each other. If $J_i$ is the flux of the $i$th diffusing substance (such as electric current density, thermal flux or particle flux), and if $X_i$ is the $i$th *generalized force* or *affinity* (or gradient) (such as the electric field, temperature gradient or pressure gradient) which produces, drives or impels the $i$th flux, then

$$J_i = L_{ij} X_j , \qquad (\text{II.1})$$

where repeated indices indicate summation. If the various diffusion processes were independent then $L_{ij}$ would be zero except when $j = i$. Onsager postulated that the diffusion processes are, in fact, coupled and the coefficients in Equation II.1 satisfy the requirement that

$$L_{ij} = L_{ji} \neq 0 , \qquad (\text{II.2})$$

(A change of sign occurs when a magnetic field is involved).

This postulate, Equation II.2, may be justified by a number of plausibility arguments involving microscopic fluctuations and time reversibility. Its validity, however, rests upon the fact that it predicts the correct experimental results for multiple diffusion processes in systems not too far from equilibrium.

It should be noted that nonlinear diffusion processes, such as exist in gaseous plasmas, can occur and that nonlinear coupling between these processes may occur, which does not agree with the Onsager assumptions, Equations II.1 and II.2.

A number of men have worked on multiple diffusion processes following Onsager, such as Prigogine (1961) and deGroot (1952). In order to generalize, the time rate of dissipation of energy may be considered. In terms of the generalized forces and the fluxes, the energy dissipated is given by the expression

$$\dot{Q} = T \, \dot{S} = J_i \, X_i > 0 \,, \qquad\qquad (II.3)$$

where the dots above the symbols refer to time rates, and the inequality comes from the second law of thermodynamics. Equation I.16, for the irreversible case. Generalizing from Rayleigh's *principle of least dissipation of energy* (1894), Onsager introduced the principle of the least entropy production $\dot{S}$ (the time rate of increase of the entropy). The entropy production may be expressed as a function of the generalized forces by substituting Equation II.1 into II.3 to yield

$$\dot{S} = (L_{ij}/T) \, X_i \, X_j \,. \qquad\qquad (II.4)$$

Differentiating Equation II.4 with respect to $X_i$ and using the Onsager relations II.2, then gives

$$\delta\dot{S}/\delta X_i = 2 \, L_{ij} \, X_j = 2 \, J_i \,. \qquad\qquad (II.5)$$

This means that $\dot{S}$, regarded as a function of the generalized forces, is a minimum for those fluxes that tend to zero, i.e. $J_i = 0$. As a system moves toward thermodynamic equilibrium the fluxes tend to zero and the entropy production toward a minimum (in particular, a zero minimum when thermodynamic equilibrium is finally reached). Similarly, as a system moves toward a steady state equilibrium where no further change with time occurs, the entropy production will tend toward a minimum (some, but not all, of the $J_i$'s in Equation II.5 being zero).

## Optimum Utilization of a Source of Energy

One of the crucial questions that irreversible thermodynamics should attempt to answer is: What are the conditions for the optimum utilization of an energy source using realistic irreversible processes and devices operating at a finite rate? Besides the value of the answers to such a question in engineering and in practical everyday life, the importance of this question in ecophysics may be judged from the fact that individual species and ecosystems which utilize an energy source optimally will displace species and ecosystems that do not. Species and ecosystems evolve toward higher efficiencies, i.e. mass supported to the energy consumed. An automobile that gives more miles to the gallon of gasoline is regarded as more desirable; as a consequence automobiles have evolved toward greater gas mileage. While such ideas may seem obvious, care must be taken to define precisely what is meant by the *optimum utilization* of an energy source.

In the previous Chapter the utility of a source of energy was defined in terms of the amount of mechanical work that it could yield in principle. The optimization of the utilization of a source of energy is, therefore, assumed to be equivalent to maximizing the amount of work obtained from the source of energy.

Most of the conditions for the optimum utilization of a source of energy may be obtained from the case of a source of thermal energy. The problem then reduces to an analysis of an irreversible heat engine operating at a finite rate.

Since a reversible heat engine (i.e. a Carnot engine) must operate infinitely slowly to remain reversible, a finite reversible heat engine will deliver mechanical power at a zero rate. The actual utility of such an ideal reversible heat engine is, consequently, zero. Even if a Carnot engine of a finite size could be actually constructed, it would have no value, since it would deliver zero mechanical power no matter how large the heat source.

PERFORMANCE MEASURED BY THE TIME RATE OF PROFIT. One of the surprising features of the present research is that reasonable answers to the question of how to optimally utilize a heat source requires the introduction of the nonphysical parameter, the time rate of profit. The time rate of profit may be appreciated as an intuitively, psychologically or economically satisfying measure for the evaluation of the performance of any device, including a heat engine.

The traditional measure to evaluate the performance of a heat engine is the *efficiency,* $\eta$, defined as the ratio of the work done W to the heat consumed Q,

$$\eta = W/Q , \tag{II.6}$$

It is assumed that the higher the efficiency the more desirable the heat engine. Not only is this traditional measure found in all introductory texts, it is also the measure most frequently used in research work to evaluate the performance of heat engines. Despite the plausibility of this measure and the fact that it is generally accepted, it does not provide an adequate measure for the evaluation of the performance of an actual heat engine operating at a finite rate with irreversible heat losses. Basically the problem lies in the fact that not only is W a function of the design parameters but Q is also a function of the design parameters, so that when $\eta$ is maximized to ascertain the optimum design parameters one may be merely minimizing the heat input instead of maximizing the work output. Measures that must be maximized or minimized which become complex functions of the parameters of the system should never be taken as ratios. For example, the optimization of an amplifier should not be based upon a maximization of the amplification, defined as strength of the signal out to the signal in.

The traditional efficiency is not a physical parameter in the sense that it is needed to obtain further physical results. It is actually an arbitrarily chosen function of W and Q that satisfies the psychological need for a measure that is larger for more work done or less energy expended. The efficiency has no inherent advantage over the time rate of profit; they are both nonphysical parameters. In actual application, however, the time rate of profit gives the desired answers, whereas the efficiency does not.

The fact that the traditional efficiency is not the best measure for evaluating the performance of a heat engine has already been noted by a number of engineers interested in the improvement of the performance of heat engines (e.g. Osterle, 1963; Kirkley, 1962; Walker, 1962). Unfortunately alternative measures that have been proposed are not universally applicable to all real heat engines and are usually defective, being arbitrarily chosen ratios. For example, Kirkley and Walker introduce five such measures (power parameters defined as dimensionless ratios) to evaluate the performance of an idealized Stirling engine. They erroneously maximize these measures to try to optimize operating conditions and design parameters, overlooking the fact that maximizing their measures is achieved in part by minimizing their arbitrarily chosen variable denominators. Only measures, such as the time rate of profit, that possess no variable denominators (which might be erroneously minimized) can be appropriately maximized.

An accurate specification of the time rate of profit, which would include such things as the amortization of the original cost of the device, will not be considered here, even though an accurate evaluation of the performance of a real heat engine would require a detailed specification of the time rate of profit. For clarity, only a simple approximation that displays the principles will be considered here. In particular, the time rate of profit $\dot{R}$ (in units of dollars per day, for example) realized when operating a heat engine is assumed to be given by

$$\dot{R} = A\,\dot{W} - B\,\dot{Q}\,, \qquad (II.7)$$

where A is the price or value per unit of work W delivered by the engine and B is the cost or value per unit of heat Q consumed by the heat engine. Dots above the symbols indicate time rates. If the price or value of work delivered A is assumed to have some standard or universal value (e.g. unity), then it is possible to analyze the performance in terms of a single parameter b, the ratio of the cost of heat consumed to the price of work delivered,

$$b = B/A\,. \qquad (II.8)$$

Substituting Equation II.8 into II.7 gives

$$r = \dot{W} - b\dot{Q}\,, \qquad (II.9)$$

where $\dot{r} = \dot{R}/A$ is the time rate of profit measured in units of power—watts, for example.

If b is specified or is related to the physical parameters, the heat engine may be evaluated in terms of the physical parameters only. Certain situations may provide this opportunity. For example, when the heat energy is supplied free, $b = 0$, and the evaluation may be made in terms of the physics only. In this case maximizing the rate of profit $\dot{r}$ is equivalent to maximizing the mechanical power output.

PERFORMANCE OF AN IDEALIZED IRREVERSIBLE HEAT ENGINE. To illustrate the principles involved, a particular idealized irreversible heat engine will be considered. An idealized device that approximates an actual heat engine more closely than a Carnot

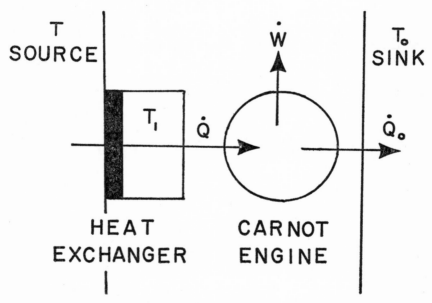

Figure 2. An idealized irreversible heat engine consisting of a Carnot engine plus a heat exchanger.

engine all by itself is a Carnot engine plus a heat exchanger as diagrammed in Figure 2. Heat flows at the rate $\dot{Q}$ from an infinite heat source at temperature T through a finite heat exchanger with finite thermal conductivity to a finite source at

Temperature $T_1$. A Carnot engine operates (it is assumed at a finite rate) between the temperature $T_1$ and an infinite sink at temperature $T_o$, absorbing heat at a rate $\dot{Q}$ from the finite source and doing work at the rate $\dot{W}$. This composite device has the following realistic features:

1. It runs at a finite rate (not infinitely slowly), and the performance depends upon the rate.
2. The size of the heat engine, as determined by the size of the heat exchanger, is limited to a fixed finite size.
3. Heat is degraded irreversibly in the heat exchanger.

This idealized device is probably the simplest model for an irreversible heat engine that can achieve these features. A heat exchanger between the Carnot engine and the heat sink would achieve a similar result.

The efficiency of the Carnot engine, as shown in Figure 2 is $(1 - T_o/T_1)$, so that the rate of doing work is given by

$$\dot{W} = (1 - T_o/T_1)\, \dot{Q} . \qquad (II.10)$$

The time rate of flow of heat through the heat exchanger will be assumed to be proportional to the temperature difference, or

$$\dot{Q} = K\,(T - T_1) , \qquad (II.11)$$

where $K$ is proportional to the size of the heat exchanger and to the thermal conductivity. Substituting Equations II.11, II.10 and II.8 into II.7 yields the time rate of profit

$$\dot{R} = AK\,T\,(1 - T_1/T)\,(1 - b - T_o/T_1) , \qquad (II.12)$$

where $T_1$ may be regarded as an adjustable parameter in this expression.

The temperature $T_1 = T^*_1$ for which the rate of profit $\dot{R}$ is a maximum may be obtained by differentiating Equation II.12 with respect to $T_1$ and setting the result equal to zero, thereby yielding

$$T^*_1 = [T_o T/(1 - b)]^{\frac{1}{2}} , \qquad (II.13)$$

which depends explicitly upon the price ratio $b$, Equation II.8. Substituting this result, Equation II.13, into Equation II.12 then yields the maximum rate of profit

$$\dot{R}_{max} = AK\,T\,[(1-b)^{\frac{1}{2}} - (T_o/T)^{\frac{1}{2}}]^2\,. \qquad (II.14)$$

This result is shown in Figure 3 as a function of the fractional temperature difference

$$\eta_c = (T - T_o)/T\,, \qquad (II.15)$$

which is also the efficiency that a Carnot engine would have operating between the infinite source at the temperature T and the infinite sink at the temperature $T_o$. The price ratio b is shown as a parameter.

If, instead of maximizing the time rate of profit $\dot{R}$, the traditional efficiency is maximized, then $T_1$ becomes equal to T and the efficiency becomes just the efficiency of a Carnot engine operating directly between the infinite source at T and the infinite sink at $T_o$. But when $T_1 = T$ the rate of energy flow to the heat engine goes to zero and the rate of delivery of mechanical work goes to zero, and, consequently, the time rate of profit goes to zero. Clearly the traditional efficiency of this heat engine is not an appropriate measure to evaluate its performance, and, therefore, in general the traditional efficiency is not an appropriate measure to evaluate the performance of heat engines.

To further stress the insufficiency of the traditional efficiency as a measure to evaluate the performance of a heat engine, the time rate of profit, Equation II.12 may be expressed as a function of the traditional efficiency $\eta$, where from Equations II.6 and II.10

$$\eta = 1 - T_o/T_1\,, \qquad (II.16)$$

by eliminating $T_1$; thus,

$$\dot{R} = AK\,T\,(\eta_c - \eta)\,(\eta - b)/(1 - \eta)\,, \qquad (II.17)$$

where $\eta_c$ is defined by Equation II.15. This result is shown in Figure 4 (for one particular choice of b and $\eta_c$). It may be noted that the time rate of profit and the traditional efficiency are reciprocally related for larger values of the efficiency, $\dot{R}$ eventually going to zero when $\eta = \eta_c$.

The expression for the maximum time rate of profit, $\dot{R}_{max}$, Equation II.14 immediately suggests ways in which the performance of the heat engine might be improved. In particular, if K

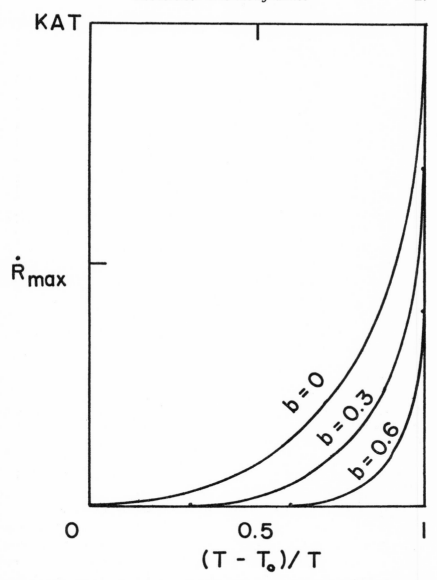

Figure 3. The maximum rate of profit, $R_{max}$, Equation II.14, as a function of the fractional temperature difference between source and the sink for the heat engine indicated in Figure 2.

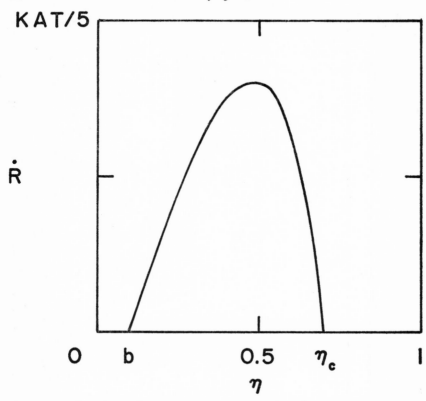

Figure 4. The relationship between the rate of profit $\dot{R}$, Equation II.17, and the traditional efficiency, $\eta$, for the irreversible heat engine indicated in Figure 2 for the case $b = 0.1$ and $\eta_c = 0.7$.

is increased the rate of profit will increase. This means that the heat exchanger should be as large as possible and should be constructed of materials with the greatest thermal conductivity. The traditional efficiency for this idealized device, Equation II.16, provides no such insight. The time rate of profit will also yield insight into the optimal design and operational parameters for heat engines in general, while the traditional efficiency will not.

A POSSIBLE PHYSICAL MEASURE FOR THE TIME RATE OF PROFIT. In principle it is possible to circumvent the introduction of the nonphysical time rate of profit if the entire ecosystem is considered instead of just the heat engine by itself. In particular, some place in the ecosystem a fraction of the work produced by the heat engine has to be fed back (by whatever devious route) into the system to help procure the heat (since in general it is not free). The fraction of the work fed back into procuring heat may be assumed (ideally) to have a value equal to the value of the heat procured bQ; or

$$\dot{W}_{feedback} = b\dot{Q} . \qquad (II.18)$$

The net rate that work is available for creating thermodynamic order in the ecosystem is then

$$\dot{W}_{net} = \dot{W}_{engine} - \dot{W}_{feedback} . \qquad (II.19)$$

Comparing this result, Equation II.19, with Equation II.7 or II.9, it is apparent that the time rate of profit is proportional to the time rate that mechanical work is made available for creating thermodynamic order in the ecosystem,

$$\dot{R} = A \dot{W}_{net} . \qquad (II.20)$$

It is now seen that the desire to maximize the time rate of profit when operating a heat engine is actually equivalent to a desire to maximize the time rate that thermodynamic ordering processes can be made available to the ecosystem. See the work of H.T. Odum (1971).

### Conditions for Ordering Processes to Occur

Another crucial question in irreversible thermodynamics is: What are the conditions that give rise to entropy reducing processes in real irreversible systems? This question is of great importance here, since one of the most significant characteristics of living systems is their ability to create and maintain an internal state of low entropy (*see* Morowitz, 1968). Since physical systems when left to themselves (i.e. isolated) proceed from a state of thermodynamic order or low entropy to a state of thermodynamic disorder or high entropy, only very special conditions are compatible with thermodynamic processes in the reverse direction.

Systems that decrease in entropy with time cannot be closed

systems, otherwise they would violate the second law of thermodynamics. Therefore, thermodynamic ordering processes can occur only in open thermodynamic systems that are exchanging
energy and/or matter with their environment. A steady state
situation is assumed here; for example, an isolated heterogeneous
system which may support a localized transient decrease in entropy is not of interest.

From the second law of thermodynamics, Equation I.16, it is
apparent that in order for a system to decrease in entropy, the
remainder of the universe must suffer a correspondingly larger
increase in entropy; thus,

$$( - dS)(\text{internal to system})$$
$$\leqq$$
$$dS(\text{universe external to system}) . \qquad (\text{II.21})$$

An important conclusion may be drawn from Equation II.21:
The larger the right side, the larger the left side can be potentially. Thus the larger the increase in the entropy external to the
system, the greater the potential the system has for creating
thermodynamic order internally. Ordering processes, including
life, can only arise in environments that are increasing the entropy of the universe with time. The environment that produces
the greater time rate of increase in the entropy of the universe
(the entropy production) will provide the greater potential for
ordering processes to occur. The word *potential* is used here to
indicate that entropy production is a necessary condition for an
environment to support entropy reducing processes, but is not
a sufficient condition. Thus, the word is used in precisely the
same sense as the word *potential* in *potential energy.*

From the definition of entropy, Equation I.15, entropy production implies a dissipation of energy from a state of high utility (large T) to a state of low utility (low T). It is, thus, only
by the dissipation of energy that a system can create an internal
state of low entropy. For example, green plants dissipate high
utility solar energy by converting it into low utility thermal energy (neglecting the small fraction that becomes stored as chemical energy). In this way green plants convert gaseous $CO_2$ and
liquid $H_2O$, which have high entropy, into solid cellulose which

has low entropy. They also reduce the entropy of the atmosphere when the small concentrations of $CO_2$ is separated from the remaining constituents of the atmosphere (the entropy of mixing).

Finally, ordering processes can occur only if there are raw materials available to be ordered. The appropriate atoms or molecules must be transported to the system that is doing the ordering. The system must also have internal mechanisms for mass transport, so that each atom or molecule can be led to its appropriate site.

Summarizing, in order for thermodynamic ordering processes to occur, the four following conditions must be present:

1.  open thermodynamic systems,
2.  entropy production,
3.  sources of energy, and
4.  mass transport mechanisms.

## A Simple Example of Steady State Entropy Production

Considering the present simple example of a system which has a steady state entropy production helps to illustrate the essential thermodynamic properties of an ecosystem. The system is composed of an ideal gas contained between a plane boundary surface at $x = -b/2$ maintained at the temperature $T_o + \triangle T$ and a plane boundary surface at $x = b/2$ maintained at the temperature $T_o - \triangle T$.

When the temperature difference is zero, $\triangle T = 0$, the entropy $S_o$ per unit area of the boundary surfaces, as given by the ideal gas formula (Sage, 1965), is

$$S_o = C_v \ln T_o , \qquad (II.22)$$

to within an additive constant where $C_v$ is the heat capacity at constant volume per unit boundary area. When the temperature difference is no longer zero the entropy per unit volume is of the form

$$dS/dx = (C_v/b) \ln T(x) . \qquad (II.23)$$

Assuming the thermal conductivity of the gas does not change over the temperature range of interest, the steady state tempera-

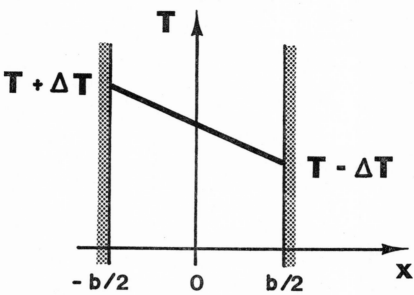

Figure 5.

ture will be a linear function of the distance between the boundary surfaces, so that the entropy per unit volume becomes

$$(dS/dx)_t = (C_v/b) \ln (T_o - 2x \triangle T/b) . \qquad (II.24)$$

Integrating with respect to x yields the total entropy per unit area

$$S_t = C_v \ln T_o - C_v\{1 + (1/2y) \ln [(1-y)/(1+y)] \\ - \tfrac{1}{2} \ln (1-y^2)\} (II.25)$$

$$= C_v \ln T_o - C_v\left[ \frac{y^2}{2.3} + \frac{y^4}{4.5} + \frac{y^6}{6.7} \right] + \cdots,$$

where $y = \triangle T/T_o$ .

The rate of energy flow per unit area is given by

$$dQ/dt = 2 \sigma \triangle T/b , \qquad (II.26)$$

where $\sigma$ is the thermal conductivity of the gas. The entropy production per area may be computed by noting that the heat source on the left experiences a decrease in entropy per unit area of the amount $(T_o + \triangle T)^{-1}$ dQ/dt, while the heat sink on the right experiences an increase in entropy per unit area of the amount $(T_o - \triangle T)^{-1}$ dQ/dt. The net entropy production by the gas per unit area is then

$$dS_t/dt = (4\,\sigma/b)\,y^2/(1 - y^2)\,, \qquad (\text{II.27})$$

where again $y = \triangle T/T_o$.

This simple steady state example illustrates some of the essential thermodynamic features of systems that can support entropy reducing processes. It is an open thermodynamic system exchanging energy with the environment. By virtue of the way the boundary conditions have been chosen, the internal energy per unit area, which is equal to $C_vT_o$, remains the same for all values of $\triangle T$. The mass per unit area also remains fixed. When the entropy production, Equation II.27, is increased from zero (i.e. $\triangle T$ is increased from zero), thermodynamic order is created and maintained (i.e. the entropy decreases and remains less, Equation II.25). The greater the entropy production, the greater the amount of order created and maintained. To create and maintain this thermodynamic order energy is dissipated, Equation II.26. The greater the rate of energy dissipation, the greater the amount of order that can be created and maintained.

Over geological times the earth has not remained in precise steady state equilibrium. Ecosystems have evolved with time. Therefore, it is of interest to inspect a simple example of an entropy producing system that experiences an explicit transient change with time.

Considering the example above where the temperature is now assumed to vary with time, the appropriate equation for the temperature as a function of time and position is the one-dimensional diffusion equation,

$$\partial T/\partial t - k\,\partial^2 T/\partial x^2 = 0, \qquad (\text{II.28})$$

where k is the diffusivity (Ingersoll *et al.*, 1948) defined by $k = \sigma/\rho c$ where $\rho$ is the density, c is the specific heat at constant volume per unit mass, and $\sigma$ is the thermal conductivity.

If there is an initial state of steady state equilibrium for a temperature difference $2 \triangle T$ and if the boundary temperatures are suddenly changed to $T_0 + \triangle T_1$ on the left and $T_0 - \triangle T_1$ on the right, the appropriate initial and boundary conditions may be expressed as

$$T = T_0 - 2x \triangle T/b + (\triangle T_1 - \triangle T) [1$$
$$- u(x+b/2) - u(x-b/2)] , \qquad (II.29)$$

where $u(x)$ is the unit step function, zero for $x < 0$ and unity for $x \geqq 0$. The solution of Equation II.28 subject to Equation II.29 may be taken as the final linear steady state solution from $T_0 + \triangle T_1$ to $T_0 - T_1$ plus an appropriate transient solution that vanishes for $t \to \infty$. Assuming that T may be written as the product of a time function and a space function and expanding the space function into an odd Fourier series (Morse and Feshbach, 1953), the solution is found to be

$$T = T_0 - 2x \triangle T_1/b + (2/\pi)\sum_{n=1}^{\infty} \{2(-1)^n \triangle T$$

$$+ (\triangle T_1 - \triangle T)[(-1)^n - \cos(n \pi/2)]$$

$$\}n^{-1} \sin(n \pi x/b) \exp [- (kn^2 \pi^2/b^2)t] . \qquad (II.30)$$

The important characteristic of this transient solution, Equation II.30 is the factor with the exponential time decay. Whether or not the new temperature difference $\triangle T_1$ is greater than or less than the original temperature difference $\triangle T$, the solution proceeds monotonically in time to its final steady state temperature distribution. If the temperature difference is increased the entropy of the gas decreases, and if the temperature difference is decreased the entropy of the gas increases. The monotonic character of the time variation of the entropy, as found for this particular example, may be anticipated in general for any non-isolated system proceeding toward a new steady state equilibrium, since it may be shown that under the present circumstances the steady state equilibrium is a state of stable equilibrium (Prigogine, 1961). If the surface of the earth has been proceeding toward a state of higher thermodynamic order in the past, it will then continue to do so in the future until a final steady state equilibrium is reached (assuming no change in the solar flux).

## Origin of Structures Differentiated in Space and Time

With sufficiently large generalized forces, gradients or affinities, ordered structures begin to appear, such as convection cells, vortices or chemical densities differentiated in space and time. Carrying irreversible thermodynamics the next step beyond the linear or quasireversible theory involves a prediction of the conditions necessary for the onset of such structures. Prigogine and Nicolis (*see* Glansdorff and Prigogine, 1971) have been carrying on extensive research in this area. Life as seen today is highly structured. Such structures had to evolve from less structured or completely nonstructured origins. The study of the dynamic origin of structured systems is, thus, pertinent to the study of life and ecophysics. The present volume, however, does not pursue the subject further.

# CHAPTER III

# AN ECOPHYSICAL DEFINITION OF LIFE AND LIVING*

THE STUDY OF ECOPHYSICS involves the application of the knowledge of physics to the study of ecology; and the study of ecology is the study of living organisms, their interrelationships and their relationships to the environment. Consequently, an appropriate definition of *life* must be chosen here.

### Basic Requirements of a Definition

A definition in general should be as explicit as possible, as simple as possible, as broad as possible, and have as great a utility as possible. The key word here is utility. The definition desired here should have the greatest utility for the study of ecophysics.

In order to be compatible with the *physics* in *ecophysics,* the definition of the property *life* should apply to a physical system using only physical parameters. In order to be compatible with the *eco* in *ecophysics,* the definition of the property *life* should depend only upon the nature of the interactions of the physical system with the rest of the universe. A living system is, thus, to be viewed as a *black box.* The interest here is largely focused

---

* Some of the basic ideas presented in this chapter and the next one were originally presented in a paper (Wesley, 1967).

on what a system does rather than on the internal nature of a system. This point of view is precisely that of classical thermodynamics, as opposed to a statistical-mechanical point of view.

## Some Significant Thermodynamic and Ecological Observations of Life

The observation of ordinary living organisms around us reveals them to be open thermodynamic systems which exchange both matter and energy with the environment. Such organisms are composed of atoms that were originally in the environment. The entropy of these atoms in the organisms is lower than the entropy of these same atoms when they were in the environment (Schroedinger, 1956; Asimov, 1962; Pardee and Ingraham, 1960; Prigogine, 1969).

Autotrophic organisms such as green plants under the action of sunlight convert $CO_2$ and $H_2O$ of high entropy into $O_2$ of high entropy and compounds of low entropy, such as cellulose, which are deposited internally. Considering this process alone, a net reduction in the entropy of the surface of the earth is produced. Most of the low entropic compounds deposited internally in autotrophs eventually become recycled into the environment. Plant respiration returns some as $CO_2$ and $H_2O$ directly to the environment. Some of these compounds are eaten by heterotrophs and subsequently oxidized. Some are returned to the environment in brush and forest fires. Some are deposited in detritus and are subsequently oxidized by detritus feeders or by slow atmospheric oxidation. The cycling of compounds in the ecosphere is not quite perfect, however, and there is an important secular decrease in the entropy of the earth's surface over geologic times (as will be discussed more thoroughly in the next Chapter).

Heterotrophic organisms are also open thermodynamic systems which exchange matter and energy with the environment. They are also composed of atoms that were originally in the environment, and these atoms are in a lower state of entropy internal to the organism as compared with the entropy of these same atoms when they were in the environment. In contrast to autotrophs, however, heterotrophic organisms, in lieu of sunlight, convert compounds at low entropy into compounds of high entropy which are then returned to the environment. This process by itself tends to increase the entropy of the earth's surface. However, considering the quasi-steady state cycling of compounds in the ecosphere, heterotrophs apparently do contribute to the secular decrease in the entropy of the earth's surface over geologic times.

## An Ecophysical Definition of Life

Considering the observations discussed in the previous section, the following ecophysical definition of life is proposed here: *Life* is taken to be a property of a thermodynamic system which

1. is open, exchanging energy and/or matter with the environment,
2. is composed of atoms whose entropy is lower in the system than the entropy of the same atoms when they were originally in the environment, and
3. is part of the overall secular process of entropy reduction of the ecosphere over geologic times.

This ecophysical definition of life is proposed primarily as a descriptive guide; it cannot be regarded as being completely satisfactory. For example, it may sometimes be difficult to decide whether or not a given system actually plays a role in the general entropy balance of the ecosphere. Similarly, it is not clear just precisely how low an entropy the internal state of a system should have in order to qualify as living. Because of the crystalline-type of arrays revealed in organelles by electron microscopes, it will be assumed here that life is a property of systems which are essentially in a solid state or are very close to it.

## The Everyday Concept of Life and the Ecophysical Definition

A single definition of *life* that fits all the ways the word is ordinarily used is clearly impossible. The choice of an adequate definition is further confounded by the rich emotional aura surrounding the word. Many of the meanings apply to an inner, personal or spiritual life, which need not be of concern here.

Apart from the inner self the word *life* in everyday situations is used anthropomorphically. A stone is not alive because it does not move, grow or appear to do anything. Plants are alive because they grow, like a man, and they can be killed, like a man. Animals that move are more lifelike than plants that do not move. Vertebrates are more lifelike than annelids. An intelligent chimpanzee is more likelife than an unintelligent perch, etc.

The everyday concept of life is based upon macroscopic observations of the behavior of macroscopic objects. The everyday concept of life does not depend upon a knowledge of cells and chemistry. In this respect the present ecophysical definition of life is compatible with everyday usage, since the definition is based primarily upon what a system does and not upon its particular internal structure or chemistry.

To stress this point, the following two hypothetical examples may be considered. First, a clever manufacturer has managed to produce robots made of steel, brass, plastic, etc., which move, act and think like humans. Unless dissected they are indistinguishable from humans. Such robots, being indistinguishable from humans, would be regarded as humans and would, therefore, be regarded as alive. The everyday concept of life, like the ecophysical definition, depends upon the dynamical behavior of an object and not upon the particular internal structure and chemistry. Second, a space man lands on Mars and, upon leaving his space ship, is immediately attacked by a seven-armed, purple Martian. To protect himself he kills the Martian with his laser gun. The behavior of the object revealed itself as alive in both the everyday sense and in the ecophysical sense, no matter what the nature of the internal structure.

It is not merely an aberration of human language that the engine of an automobile which is not running is *dead*, a television set that is turned on comes to *life*, or a machine is sometimes *killed* when it is stopped. These are, in fact, accurate applications of the everyday concept of life which are equally valid in the ecophysical sense.

The present ecophysical definition of life probably comes closer to the real essence of what is usually meant by life than any other scientific definition as yet proposed.

*Some Restricted Scientific Definitions of Life*

From time to time a variety of attributes have been proposed in an attempt to establish a scientific definition of life. Following is a partial list of attributes that have been frequently proposed:

1. *Carbon chemistry*: The expression *organic chemistry* arose almost two hundred years ago because of the original belief that such chemistry was associated only with life. Today a more general designation is given simply by *carbon chemistry*. But there is still a widely held belief that life must necessarily be based upon carbon chemistry. This belief then constitutes a defining property of life.

2. *Self-replication*: Many biologists (e.g. Fox, 1971) prefer to add self-replication as a defining criterion for life. For example, geneticists prefer to think of life as a self-replicating system that carries the information necessary for its own self-replication. While crystals usually grow by self-replication, this criterion would appear to rule out the possibility of ascribing life to machines.

3. *Death*: Apart from the germ cells or seeds, it is frequently assumed that every organism must eventually die in order for reproduction and evolution to take place.

4. *Genetic code*: It is frequently assumed that a unique aspect of life is the information carried by living things which gets transmitted to future generations, determining the nature of succeeding generations. A defining property of life is then taken to be the existence of a genetic code which permits self replication.

5. *Capacity to evolve*: The capacity of a species line to evolve is frequently taken to be a defining property of life. It is assumed that random changes in the genetic code produces random changes in the character of individual organisms. Natural selection is then assumed to select only those individuals with the highest survival potential.

6. *Protein and amino acids*: Since proteins and amino acids are the essential building materials for ordinary carbon-based life, it is sometimes suggested that the presence of protein and

amino acids helps to differentiate the living from the nonliving (e.g. Schwartz, 1971).

7. *Optical activity*: Since the proteins found in ordinary carbon-based life possess a particular helical symmetry, it has been suggested (Stryer, 1966) that the search for extraterrestrial life should involve a search for optically active proteins.

8. *DNA*: Since it appears that the genetic code for ordinary carbon-based life is registered on DNA molecules, it has been proposed that the basic essential of life is DNA.

9. *Metabolism*: It is generally felt that the definition of life should take cognizance of the dynamic nature of life. The utilization of energy and food to build structures and to carry on various processes has been often regarded as one of the essential aspects of life.

10. *Enzymes*: On occasion it has been suggested that the basic living process is primarily an enzymatic process.

11. *Cells and membranes*: Since the original discovery of the cellular nature of most ordinary carbon-based life, it has been generally assumed that life must consist of cells. In particular, it is assumed that the highly selective functions of membranes are necessary for life (e.g. Oparin, 1971).

12. *Reaction to stimuli*: It is frequently felt that a distinguishing property of life is its ability to respond to stimuli. It is generally assumed that the response enhances the survival of the organism. While any physical system will respond somewhat to almost any physical stimulus, the response of the nonliving system is generally quite small compared to that of living systems.

These attributes are all compatible with the general ecophysical definition proposed above. The ecophysical definition, thus, presents a very broad definition, while the items in the list above give rise to more restricted definitions of life. While some of these restricted definitions may be of value in some limited areas of research—such as genetics, the physiology of carbon-based life, and the evolution of carbon-based life—the study of ecophysics, involving the study of physical systems of low internal entropy that play an active role in the ecosphere, need not involve such restrictive attributes (except possibly 3, 9 and 12 above).

### The Utility of the Ecophysical Definition of Life

Thermodynamic principles are among the most general, basic and fruitful of scientific principles. They are applicable to both simple and complex systems alike. The present definition of life, being based on these principles, is about as broad as possible and should prove to be scientifically fruitful.

The relationship of life to the environment is, above all, a thermodynamic relationship. Consequently, a definition of life based on the thermodynamic role that life plays in the environment should prove fruitful in the study of ecology.

The definition proposed here is of sufficient generality to indicate the possibility of extremely alien life forms actually existing. No matter what the morphological or chemical features of such alien life, it will still be governed by precisely the same thermodynamic principles that govern life on earth. For example, a robot must have the same thermodynamic needs and the same thermodynamic behavior as ordinary carbon-based life.

The ecophysical definition stresses the dynamic behavior of living systems. It indicates what a living system is likely to do and how it is likely to behave, no matter the internal detailed structure of the system. The thermodynamic needs of man and all other life forms are basically the same. Wherever man may go and whatever alien life forms he may encounter, the thermodynamic behavior of life will always be basically predictable.

The likelihood of finding life in a particular environment depends upon the thermodynamic properties of the environment. In particular, life can evolve only in an environment which is continually increasing the entropy of the universe with time (as will be discussed in greater detail in the next Chapter).

The proposed definition indicates that the entropy of compounds taken from the environment must be reduced before being incorporated in the living system. Consequently, energy must be degraded from a state of high utility to a state of low utility. In other words, the definition indicates that life needs a supply of energy for its existence.

The thermodynamic role of machines in the ecosphere is becoming increasingly important. For example, it may be noted that the burning of fossil fuels, which includes a large fraction consumed by machines, has accounted for a 13 percent increase in the $CO_2$ content of the atmosphere in the last fifty years (Sawyer, 1972). It is relatively easy to conceive of a future technology where all of man's functions have been taken over by machines and robots. Such a system of machines is compatible with the above ecophysical definition of life. Such machine life need not depend upon carbon chemistry for structure or for energy, since solar energy and nuclear fuel may be used to power the system.

Since machines may be included in the present ecophysical concept of life, it is possible to study the coexistence of machines and man, machines being treated as any other competitive species (*see* Chapter IX for a more complete discussion). Any predictions concerning the future of the earth's biosphere would be completely unrealistic without, thus, taking into account machines as a significant form of ecophysical life. Machines displace other forms of life. They take up space. They pollute the air, land, streams, rivers, lakes and oceans with their effluvium. The proposed ecophysical definition of life allows a more realistic treatment of the relationship between machines and all other forms of life.

## Death, Reproduction and Evolution of Ecophysical Life

Death, reproduction and evolution are features of ecophysical life in general. Each individual organism or living system will for one reason or another eventually cease to function due to accidents, deterioration or planned obsolescence. When an individual system ceases to function it may be said to *die.*

A viable ecosystem must provide for the replacement of living systems that have died. In other words, ecophysical life must in general involve reproduction. While ordinary carbon-based life self replicates, other forms of ecophysical life do not. Viruses, for example, furnish their own blueprints and induce a host cell to reproduce more virus particles. Automobiles and other machines are reproduced in factories that generally retain blueprints for their manufacture.

Not only has carbon-based life evolved (over the last 4 billion years), but machine life has also evolved (over the last 4 thousand years). The process of natural selection is the same for both life forms, but the, mechanisms for the production of deviants is drastically different in the two cases (*see* Chapters VIII and IX for a discussion of evolution).

## Systems Sometimes Regarded As Living Which Are Ecophysically Dead

While the ecophysical definition of life is very broad, there is one situation where a system, which is sometimes regarded as living, is ecophysically dead. In particular, a system that is thermodynamically isolated from its environment cannot be ecophysically alive. A frozen amoeba, for example, which exchanges neither matter nor energy with its environment is, according to the ecophysical definition, nonliving. However, such a frozen amoeba has the potential for life, since it may be thawed out to again become an open thermodynamic system capable of carrying on ordinary life processes. Similarly, some spores may be ecophysically dead for all practical purposes.

## Complexity, Order and Life

It is sometimes felt that a defining property of life is its complexity. For example, it is sometimes claimed that life must be based on carbon chemistry because only carbon chemistry can provide the requisite complexity. This view, however, is in error. Carbon-based life and ecophysical life in general are, in fact, simpler systems derived from a complex environment.

The error arises from a misconception of what is *complex* and what is *simple*. In order to adequately describe natural phenomena the word *complex* must be used in its rigorous unambiguous statistical mechanical sense. The precise state of a system is specified when the momenta and positions of all of the atoms of the system are known. The complexity of a system may then be taken as the number of independent facts or statements necessary to completely specify the system. A mole of a pure substance in a gaseous phase is, thus, complex, because the position and momentum of each molecule must be specified independently. In this case $6N_0$ (where $N_0 = 6.02 \times 10^{23}$ is Avagadro's number) statements are needed. A perfect crystal of the same substance, on the other hand, is very simple (ignoring small vibrations), since the precise state of the crystal may be specified by the lattice spacings, the position of the crystal, and the orientation of the crystal—in this case no more than about nine to twenty-four statements. The entropy also measures the complexity of a system, the greater the entropy the greater the complexity. The *order* of a system may be measured by an appropriate large constant minus the entropy (or order may be measured by the negentropy). Thus, a gas is complex, has high entropy, and is disordered, while a perfect crystal is simple, has low entropy, and is ordered. Naïvely one may be inclined to view a gas as simple, since it would appear to require only three parameters (the temperature, pressure and density) to specify its state. This view, however, is grossly in error (*see* the discussion in Chapter I.)

Since life is composed of atoms in a lower entropy state than the entropy of the same atoms when they were in the environment, life is simpler than the complex environment from which it came. A correct view of life should be associated with simplicity and not complexity.

### Growing Crystals, the Simplest Form of Ecophysical Life

A salt crystal growing in a saturated solution fits the present criteria for ecophysical life, assuming that it actually functions in the ecosphere. Such a crystal is an open thermodynamic system in the process of exchanging matter and energy with the environment. The ions in solution are in a disordered array with a high entropy. These ions then become deposited in the crystal in a highly ordered array of low entropy. In order to indicate that attributing life to growing crystals may not be completely unreasonable, a number of points may be made:

Electron microscopes are revealing the fact that ordinary life possesses many essentially crystalline features. Organelles are shown to possess orderly arrays of complex molecules (e.g. Fischman and Weinbaum, 1967). This suggests the possibility that ordinary carbon-based life may be basically crystalline in nature, which may be contrasted with the concept in the first third of this century that life was essentially a solution or soup called *protoplasm.*

Growing crystals, like any form of ecophysical life, can play a significant role in the thermodynamics of the surface of the earth and can, therefore, interact significantly with more ordinary carbon-based life forms. For example, the crystallization of water on the earth is well known to play a significant role in the patterns of life on the earth.

It is significant that crystals can grow only in environments that permit entropy-reducing processes to occur, as is true of all living processes. It is conceivable (and may actually occur) that growing crystals can compete with ordinary carbon-based life for available energy and compounds.

Crystals of various compounds have been deposited in the earth as low entropic ore bodies. These deposits have been utilized by man and machines. Within this context, growing crystals may be regarded as autotrophic organisms whose dead bodies provide food for advanced heterotrophs. Coal is a particularly important example of such a low entropic ore.

Under ordinary circumstances in nature it appears that crystals are not spontaneously generated. Like ordinary forms of life, a seed, nucleus, germ or parent crystal is needed for the crystal to reproduce. For example, in World War I the Americans were initially unable to synthesize TNT (trinitrotoluene) in crystalline form; it remained in a liquid phase. It was only after crystalline TNT from a dud German shell was imported into the United States that it became possible to manufacture crystalline TNT. Today there are enough nuclei in the air at all times to produce crystalline TNT with ease. While it is possible in the laboratory to get crystal growth on a seed of different material (which must usually have a similar structure), it is frequently difficult, if not impossible, to achieve. Such phenomena seldom occurs in nature.

High utility solar energy, being dissipated to thermal energy, has provided the potential for ordering processes on the surface of the earth since the earth first cooled down to its present state almost 4½ billion years ago. It is possible to speculate that the earliest forms of life were growing crystals. Early evolution may have been simply the competitive crystallization (and/ or polymerization) of amino acids (Cairns-Smith, 1971).

## The Need for a Continuous Measure of Living

Any phenomenon having both dichotomous and continuous variables associated with it is more likely to be understood satisfactorily by concentrating on the continuous variables in preference to the dichotomous variables (*see* the beginning of Chapter XI for further discussion on this point). Thus, the dichotomy between life and not-life (or death) or animate and inanimate is probably best generalized to a continuous scale. Systems can then take on aspects of being more alive, less alive, inanimate, etc. The word proposed here for such a continuous measure is *living*. Other words such as *livingness, vitalness* or *animateness* might also have been chosen. The dichotomous view of life or not—life may still be preserved by breaking the continuous living scale into two parts, smaller values (possibly negative) indicating a lack of life and larger values indicating the presence of life.

According to the ecophysical definition of life proposed above, the living property should involve an interaction of the living system with the rest of the universe. This implies processes that take place in time. In particular, life or living is assumed to be an on-going process. If the process stops the living stops. Since the time rates of change of continuous variables and fluxes of continuous quantities are also continuous, a continuous concept of living should be a function of the flux of energy, matter, or other such quantities into and out of the system.

## A Measure of Living

From the above discussion living is interpreted to mean, first of all, a maintenance of an internal state of low entropy or high order. This concept involves a property independent of time, a static or steady state concept. The second feature involves the rate or degree to which a system is coupled with the rest of the ecosystem. A measure of living L is then taken to be of the form

$$L = (\text{Static order}) \times (\text{Rate of involvement}) . \quad (\text{III.1})$$

A measure of the degree or static order maintained internally may be taken as the fractional difference between the entropy maintained internally $S_i$ as compared with some standard high entropy state of the same atoms $S_o$ (for example, the entropy of the same atoms as they normally occur in the environment); thus,

$$(\text{Static order}) = 1 - S_i/S_o . \qquad (\text{III.2})$$

The rate of involvement with the ecosystem cannot be as satisfactorily chosen. For the present purposes, however, it is probably sufficient to choose the energy flux through the system R per unit mass M (the carbon mass being chosen for carbon-based life) of the system which is associated with the ecosystem; thus,

$$(\text{Rate of involvement}) = R/M . \qquad (\text{III.3})$$

This rate of involvement becomes just the metabolic rate for carbon-based life.

From Equations III.1, III.2 and III.3 the measure of living then becomes

$$L = (1 - S_i/S_o)(R/M) . \qquad (\text{III.4})$$

This particular measure of living is, of course, arbitrary and not entirely satisfactory; yet it has a number of useful features. It replaces a generally qualitative concept with a concrete definition involving measurable parameters. Being measurable, L should have some scientific applications. It replaces a primitive dichotomous idea of life with a continuous concept of living. If desired the dichotomous idea of life can be recovered by choosing $L > 0$ to mean life and $L \leqq 0$ to mean the absence of life.

### Does Entropy Reduction Make Life Special or Unique?

Since life is based upon entropy reducing processes, the question arises as to how this can occur and still satisfy the second law of thermodynamics, Equation I.16. The second law of thermodynamics can be satisfied provided the reduction in entropy achieved in life is less than a corresponding increase in the entropy of the environment; or

$$- dS(\text{internal to organism}) < dS(\text{environment} \\ \text{external to organism}), \qquad (\text{III.5})$$

as already indicated by Equation II.21. The organism referred to here may be regarded as an autotroph. The equality in Equation III.5 has been omitted here, since only real irreversible processes are being considered.

Any isolated system left to itself will experience only entropy increases. Most common everyday processes involve entropy increases without any noticeable corresponding entropy decreases. The usual feeling is that entropy reducing processes are rare. The basic thermodynamic role of life is then frequently viewed as being special or unique. However, such a view is not warranted, since the entire observable universe appears to be slowly decreasing in entropy. Each star, as well as any associated planetary system, evolves from an interstellar cloud of gas and dust of high entropy through a long sequence of stages of progressively less entropy and greater order. The second law of thermodynamics is not violated, because each star radiates off into space its excess entropy in the form of electromagnetic radiation. A real paradox (as yet not satisfactorily resolved) arises only when it is asked why an equilibrium quantity of this electromagnetic radiation is not observed (*see* the beginning of Chapter VIII for a more detailed discussion).

The sun, like all other stars, is decreasing in entropy. The solar system and the earth's surface are decreasing in entropy. Life is part of this general entropy reduction of the observable universe; thus entropy reduction does not make life special or unique.

# CHAPTER IV

# ENVIRONMENTAL CONDITIONS NECESSARY FOR LIFE

THE GENERAL ENVIRONMENTAL CONDITIONS necessary for life may be deduced from the definition of life presented in the previous Chapter. The condition that life be composed of atoms in a state of lower entropy than the same atoms when in the environment requires an environment that can support entropy reducing processes. The two primary characteristics of such environments are (1) they experience an entropy production, and (2) they possess mechanisms for mass transport. The general theory of life-supporting environments as developed here is then used to estimate the likelihood of life in the solar system. Chapter V continues with the problem of where life might exist by surveying possible energy sources in the solar system.

### Entropy Production — A Measure of the Potential for Life

An important conclusion may be drawn from Equation III.5 (or Equation II.21): the larger the right side, the larger the left side can be potentially. Thus, the larger the increase in entropy external to the organism, the greater the potential the organism has for creating thermodynamic order internally. Ordering processes, including life, must take place in time, so that life can be supported only in environments that are increasing the entropy

50

of the universe with time, i.e. have an entropy production. The environment that has a greater entropy production, dS/dt, will provide a greater potential $\phi$ for ordering processes such as life to occur; thus,

$$\phi = dS/dt .\qquad\qquad (IV.1)$$

This measures not only the likelihood of life existing (apart from other necessary conditions); it also indicates the amount of life or living that an environment could sustain.

The word "potential" is used here to indicate that entropy production is a necessary condition for an environment to support life, but it is not a sufficient condition. Many environments, such as the surface of the moon and Mercury, have a large entropy production per unit area, but support no life (in the case of the moon and Mercury it is due to a lack of mass transport mechanisms.) The word "potential" used here has precisely the same meaning as in the expression "potential energy." A rock on a hillside has potential energy; it has the potential for doing work when lowered to the valley below. However, it may never actually be lowered to the valley below, so that, while the potential energy is a necessary condition for the rock to do work, it is not a sufficient condition.

The presence of mass transport mechanisms is also a necessary feature of most life-supporting environments. It is, however, of secondary importance compared with entropy production, since mass transport cannot take place in the absence of entropy production.

### Entropy on the Surface of a Planet

The surface of a planet is an open thermodynamic system which is supplied with high utility solar energy which is reradiated as low utility thermal energy. The surface of a planet increases the entropy of the universe at a steady rate. Radiant energy of the amount dQ leaving the surface of the sun at an effective temperature of about $T_1 = 5800°$ K carries an entropy $dQ/T_1$ (Epstein, 1937). If this energy dQ is absorbed by the surface of the planet an equivalent amount of energy must be reradiated into space, in order for the surface of the planet to maintain thermal equilibrium. If the temperature of the surface of the planet is $T_2$, then the energy dQ, which is transported from the surface into space (by direct radiation and convection plus radiation), produces an entropy flux from the surface of the planet of the amount $dQ/T_2$. The surface of the planet, thus, converts radiant energy from the sun carrying entropy of the amount $dQ/T_1$ to thermal energy carrying an amount of entropy $dQ/T_2$. The net entropy production (net increase in the entropy of the universe per unit time) created by the surface of the planet is then

$$\frac{dS}{dt} = \frac{dQ}{dt}\left(\frac{1}{T_2} - \frac{1}{T_1}\right). \tag{IV.2}$$

If it is assumed that the time rate of lateral mass transport is limited, the total entropy production taken over the entire surface of a planet is not as significant a measure of the potential for life as the entropy production per unit area. A detailed computation of the entropy production per unit area would take into account the variation of temperature and incident solar energy flux as a function of the time of day and the latitude. However, a preliminary estimate may be obtained by considering an average surface temperature (the daytime temperatures for bodies with very low nighttime temperatures) and by taking the amount of solar energy absorbed to be $(1 - A)$ times the incident solar energy flux where A is the spherical or Bond albedo. If the solar constant is R (2.00 calories per minute per centimeter squared or $1.40 \times 10^6$ erg/sec cm$^2$) the mean entropy production per unit area of the surface of the planet by this estimate is given by

$$P = \frac{dS}{dAdt} = \frac{Rr_e^2(1-A)}{2r^2}\left(\frac{1}{T_2} - \frac{1}{T_1}\right), \tag{IV.3}$$

where r is the mean distance of the planet from the sun and $r_e$ is the mean distance of the earth from the sun (the astronomical unit). The 2 in the denominator arises from the fact that the area of a hemisphere is twice the area of the circle intercepting sunlight. Numerical values of this estimate for the planets in the solar system are presented in Table I.

### An Environment Must Supply Energy to Support Life

Since a change in entropy may be defined in terms of heat transferred divided by the absolute temperature, Equation I.15, entropy production implies energy flow in which the character of the energy changes. Since energy is conserved, the entropy production associated with a fixed energy flow is given by

$$\frac{dS}{dt} = \frac{dQ}{dt} \left( \frac{1}{T_2} - \frac{1}{T_1} \right), \qquad (IV.4)$$

where $T_2$ is the absolute temperature to be associated with the final energy state and $T_1$ is the temperature to be associated with the initial energy state. It is thus apparent that a supply of energy is required for entropy production and is, therefore, required to sustain life.

Living organisms depend upon entropy reducing processes. The maximum amount of life that can be sustained will be given by the maximum rate of entropy reduction. For a given fixed supply of energy per unit time the maximum amount of thermo-dynamic order (or entropy reduction) that can be sustained will depend upon the optimum utilization of the energy source. This will depend upon the detailed circumstances involved. In particular, the utilization of an energy source depends upon its conversion from a state of high utility to a state of low utility as discussed in Chapter I. A survey of energy sources that might sustain life is presented in the following Chapter.

## Mass Transport and the Ability of an Environment to Support Life

Entropy reducing processes, such as are required for life, require the transport of selected molecules from widely scattered locations to a single location where they are arranged into an orderly array. Mass transport mechanisms must be present in order to move the selected molecules from place to place. The mass transport mechanisms here may be placed in three categories: diffusion, convection and motility. While diffusion processes always play a role in living systems, the following sections reveal the fact that convection and motility are generally much more important. The relative inconsequence of diffusion processes in life is a fact which has not heretofore been properly appreciated.

The type of mass transport of interest here involves processes that carry molecules to an organism, so that the organism may differentially select the molecules it desires. Since the transport of interest must be relative to the organism, it might be referred to as *relative* or *differential mass* transport. For example, the mass of the earth transported about the sun would be of no interest here.

### The Ability of Mass Diffusion to Supply Nutrients and to Carry Away Wastes

The rate of mass transport that can be achieved by diffusion alone depends upon the rate of diffusion, the geometry and the distances involved. If g in $gm/cm^3$ is the concentration of nutrient (or waste) of interest which is diffusing through another medium, the flux density of nutrient (or waste) $R$ in gm/sec $cm^2$ is given by

$$R = - D \nabla g , \qquad (IV.5)$$

where D in $cm^2/sec$ is the diffusion coefficient (Longsworth, 1957).

Diffusion of molecules through a solid is extremely slow at ordinary temperatures for ordinary substances. At elevated temperatures where melting occurs the diffusion coefficient approaches that for diffusion in a liquid. Diffusion in a solid is strongly temperature dependent where D may be frequently satisfactorily represented by the function $D_0 \exp(-H/kT)$, where $D_0$ and H are constants and k is Boltzman's constant. The value of D for a solid ordinarily lies in the range

$$0 \leqq D(solid) \leqq 10^{-7} \ cm^2/sec , \qquad (IV.6)$$

where most solids found in the environment on the earth have values of D less than $10^{-10} \ cm^2/sec$. At elevated temperatures or for rare compounds and solids D may become $10^{-5} \ cm^2/sec$. In general, it is safe to neglect the diffusion of molecules through solids as far as living processes are concerned.

The diffusion of molecules through a liquid is generally much more rapid than diffusion through a solid. In particular, the diffusion coefficient in a liquid ordinarily lies in the range

$$10^{-6} \leqq D(liquid) \leqq 10^{-4} \ cm^2/sec . \qquad (IV.7)$$

The rate of diffusion through liquids is generally about $10^5$

times as rapid as through solids. For unusual compounds or for elevated temperatures where the liquid is almost in the gaseous phase the diffusion coefficient may become as high as $10^{-1}$ cm²/sec. For liquids the diffusion coefficient generally varies as the first power of the absolute temperature (Hirschfelder *et al.*, 1954).

The diffusion coefficient for gases ordinarily lies in the range

$$10^{-1} \leq D(\text{gas}) \leq 1 \text{ cm}^2/\text{sec} . \qquad (\text{IV.8})$$

Under ordinary circumstances the diffusion through gases is about $10^4$ times greater than the diffusion through liquids. In gases the diffusion coefficient varies with the temperature approximately as $T^{3/2}$ (Hirschfelder *et al.*, 1954).

As far as diffusion in solids, liquids and gases is concerned, it appears that gaseous diffusion provides the most rapid mass transport. However, it is important to note that many of the molecules on the earth that are necessary for ordinary carbon-based life do not occur in a gaseous phase, and liquid diffusion, even though it occurs at a much slower rate, is still of primary importance.

### Sources of Nutrients and Diffusion

According to Equation IV.5 a flux of material necessary for survival by diffusion alone requires the existence of a concentration gradient, and, in order to maintain such a gradient over long distances, large concentrations of nutrients (or wastes) are required. For example, in the linear case if nutrient is supplied at the minimum rate for survival $R_x$, then Equation IV.5 yields a steady state concentration of nutrient as a function of x given by

$$g = g_o + R_x x/D , \qquad (\text{IV.9})$$

where the boundary surface of the organism is at $x = 0$ and the source is to the right of the origin and $g_o$ is the concentration at the surface of the organism. If the requirement for survival is assumed to be for $R_x = 10^{-6}$ gm/cm² sec, then in a liquid where $D = 10^{-5}$ cm²/sec the concentration of nutrient at a distance of $x = 10$ cm would have to have the impossibly large value of $g = 1$ gm/cm³ even assuming $g_o = 0$. While the numerical value of $R_x$ may have been chosen too large in this hypothetical example, the value of g at 10 cm is still going to be too large for any reasonably smaller values of $R_x$. This numerical example serves to indicate that excessively large concentrations of nutrients would have to be assumed if there were to be sizeable transport of mass over any significantly large distances by diffusion alone. In general, large concentrations of nutrients are not provided in nature. It may thus be concluded that diffusion processes generally play a secondary role in the mass transport processes necessary to maintain life.

### The Effect of Geometry on Mass Transport by Diffusion

The effect of geometry on the rate of mass transport by diffusion may be investigated by considering three idealized examples involving linear geometry, circular cylinder geometry and spherical geometry (one, two and three dimensional diffusion). The living system (having a plane surface, a circular cylindrical surface or a spherical surface) is placed in an infinite medium with an initially uniform concentration of nutrient, $g = g_0$ , or with an initially zero concentration of waste material, $g = 0$. Under these circumstances plane and cylindrical geometries eventually lead to the living system perishing; the living system is eventually unable to acquire sufficient nutrient per unit time or to eliminate sufficient waste material per unit time. Under these circumstances a living system with spherical geometry may or may not be able to survive indefinitely depending upon the numerical values of the parameters involved.

### The Choice of Boundary Conditions

In order to explore the effect of geometry, it will be sufficient here to neglect possible complexities involving the rate of transfer of nutrients across semipermeable membranes. It will be sufficient to assume that the rate of ingestion of nutrient per unit area R is proportional to the concentration of the nutrient just outside the boundary surface of the organism; thus,

$$R = kg , \qquad (IV.10)$$

where k is a constant. The rate of flow of nutrient across the boundary must be matched by a rate of flow up to the boundary. The rate of flow by diffusion in the medium is given by Equation IV.5. Equating Equations IV.10 and IV.5 at the boundary surface then yields the boundary condition

$$\mathbf{n} \cdot \nabla g = bg, \qquad (IV.11)$$

where $\mathbf{n}$ is a unit normal to the surface and $b = k/D$ is a constant.

Assuming the organism is introduced into the infinite medium at time $t = 0$ when the nutrient has a uniform concentration $g_0$, the initial condition becomes

$$g = g_0 \quad \text{for} \quad t = 0. \qquad (IV.12)$$

It is further assumed that the boundary condition, Equation IV.11, does not change with time, the surface does not move with time (as might conceivably occur during growth), and that what happens to the nutrient once it has crossed the boundary surface is no longer of any concern.

An analysis of the disposal of wastes by diffusion alone can be reduced to the identical mathematical problem of the diffusion of nutrients if the appropriate boundary conditions are chosen. In particular, the rate of elimination of wastes across the boundary of an organism per unit area $R$ may be taken proportional to a constant $g_0$ (perhaps the concentration of wastes internal to the organism) minus the concentration of wastes $g$ just outside of the organism; thus,

$$R = k(g_0 - g), \qquad (IV.13)$$

where $k$ is an appropriate constant. The rate of flow of wastes in the medium is also specified by Equation IV.5. By introducing the function

$$h = g_0 - g, \qquad (IV.14)$$

the same differential expression IV.5, the same boundary condition IV.11, and the same initial condition IV.12 as for the case of the diffusion of a nutrient is obtained. The mathematical solution of $h$ is then given immediately once the problem for the diffusion of nutrients has been solved. The physical meaning and the values of the constant parameters involved will, of course, be different.

ONE DIMENSIONAL DIFFUSION OF NUTRIENT. One dimensional diffusion may be important in many situations. For example, a layer of organisms on the bottom of a stagnant pond might depend upon the downward diffusion of $CO_2$ or other nutrients from above. Tissue is frequently formed in layers, so that tissues may depend upon linear diffusion normal to such layers.

From the equation of conservation for the nutrient in the medium,

$$\nabla \cdot \mathbf{R} + \partial g/\partial t = 0 , \qquad (\text{IV.15})$$

and Equation IV.5 the diffusion equation governing the concentration g becomes

$$D \nabla^2 g - \partial g/\partial t = 0 . \qquad (\text{IV.16})$$

For the present linear case Equation IV.16 reduces to

$$D \, \partial^2 g/\partial x^2 - \partial g/\partial t = 0 . \qquad (\text{IV.17})$$

The general solution to Equation IV.17 subject to the boundary condition IV.11 at $x = 0$ and the initial condition IV.12 is given by

$$g = g_0 \left[ 1 - \text{erfc}(X) + \exp(Y^2 - X^2) \, \text{erfc}(Y) \right] , \qquad (\text{IV.18})$$

where

$$X = x/2\sqrt{Dt} , \qquad Y = x/2\sqrt{Dt} + b\sqrt{Dt} , \qquad (\text{IV.19})$$

where $\text{erfc}(X)$ is the error function complement of X (Abramowitz and Stegun, 1965).

The time rate that the plane organism (or a planar distribution of organisms or cells) ingests nutrients per unit area as a function of time is given by Equations IV.10 and IV.18; thus,

$$R = Dbg|_{x=0} = R_0 \, e^{\tau} \, \text{erfc}(\sqrt{\tau}) , \qquad (\text{IV.20})$$

where $R_0$ is a constant equal to the initial rate of ingestion per unit area at $t = 0$ and $\tau = Db^2 t$. A good approximation of R,

Equation IV.20, for all values of t, and one which is asymptotically correct for large values of t is

$$R \approx 2R_o / (\sqrt{\pi \, \tau} + \sqrt{\pi \, \tau + 4}) \, . \qquad (IV.21)$$

The behavior of R as a function of time t, Equation IV.20 or IV.21, is indicated in Figure 6.

It is apparent from Equation IV.20 or from Figure 6 that eventually the rate of ingestion of nutrient per unit area must become vanishingly small and that the organism will perish when the rate of ingestion becomes less than some minimum value $R_x$ necessary for survival. This means that a plane layer of organisms or cells will eventually die if diffusion is the only mass transport mechanism available to supply nutrients from an infinite nonreplenished medium.

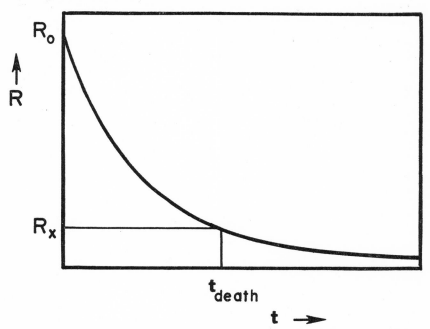

Figure 6. A diagram indicating the time rate at which a plane (or cylindrical) organism can ingest nutrient (or eliminate wastes) R from an infinite medium if diffusion processes only are present as a function of time t. Death occurs inevitably at the time $t_{death}$ when the rate of ingestion (or elimination) falls below that necessary for survival $R_x$.

Similarly a plane layer of organisms or cells will eventually die if diffusion is the only mass transport mechanism available to eliminate wastes into an infinite medium.

Two Dimensional Diffusion of Nutrients. All of the life on the surface of the earth lives essentially in a two dimensional environment and depends upon lateral mass transport for survival. The problem of two dimensional mass transport is, consequently, important. The present idealized problem of two dimensional diffusion to a circular cylindrical organism (of infinite length) should be of particular interest in the investigation of individual sessile organisms in a layer of still water.

The diffusion equation IV.16 for cylindrical coordinates is

$$\partial^2 g/\partial r^2 + r^{-1}\,\partial g/\partial r - D^{-1}\partial g/\partial t = 0 , \qquad (\text{IV.22})$$

where r is the radial distance from the center of the organism whose boundary may be taken at $r = a$. The solution of Equation IV.22 subject to the initial condition Equation IV.12 may be obtained by taking a Laplace transform of Equation IV.22 with respect to time, solving the resultant differential equation in r subject to the boundary condition IV.11 and taking the inverse transform. Changing the variable of integration from the transform variable s to the new variable of integration,

$$x = e^{-i\,\pi/2}\,a\sqrt{s/D} , \qquad (\text{IV.23})$$

yields the result

$$g = \frac{g_0}{i\pi} \int_{-\infty}^{+\infty} \frac{e^{-(Dt/a^2)x^2}}{x}\left\{1 - \frac{H_0^{(2)}(rx/a)}{H_0^{(2)}(x) + (x/ba)\,H_1^{(2)}(x)}\right\} dx ,$$

$$(\text{IV.24})$$

where the path of integration is to be taken below the real axis and $H_0^{(2)}$ and $H_1^{(2)}$ are Hankel functions of the second kind of zero and first order, respectively.

For large values of the time t most of the contribution to the integral is near the origin for $x = 0$, due to the factor exp $[-(Dt/a^2)x^2 - \ln x]$, the bracket containing the Hankel functions not being so rapidly varying. Thus, the integral may be approximated by the method of steepest descent. Expanding the exponent about its minimum value at $x = e^{-i\pi/2}a(2Dt)^{-1/2}$ to second power in a and performing the simple integration yields the approximation

$$g \approx g_0\sqrt{e/\pi}\left\{1 - \frac{K_0(rx_0/a)}{K_0(x_0) + (x_0/ba)K_1(x_0)}\right\}, \qquad (IV.25)$$

where $x_0 = a/\sqrt{2Dt}$ and $K_0$ and $K_1$ are modified Bessel functions of the first kind. For large values of t the modified Bessel functions in Equation IV.25 may be approximated for small arguments, yielding the further approximation

$$g \approx 2g_0\sqrt{e/\pi}\,\frac{\ln(r/a) + D/ka}{\ln(8Dt/a^2) - 2\gamma} + O(a^2/2Dt), \quad (IV.26)$$

where $b = k/D$ and where $\gamma$ is Euler's constant, $\gamma = 0.57722$ · · · . By setting $r = a$ and multiplying Equation IV.26 by k the rate at which the organism ingests nutrient per unit area for large times becomes

$$R \approx 2Dg_0\sqrt{e/\pi}\,/a[\ln(8Dt/a^2) - 2]. \qquad (IV.27)$$

Since eventually R goes to zero (as may be seen from Equation IV.27 or directly from IV.24), this organism must eventually die when $R = R_x$ where $R_x$ is the minimum rate of ingestion per unit area sufficient for survival. The general behavior of Equation IV.27 is also indicated in Figure 6. The larger the original concentration $g_0$, the larger the diffusion coefficient, and the smaller the radius a, the longer the organism can last. The rate of ingestion for the cylindrical case, Equation IV.27, which drops off as $(\ln t)^{-1}$, does not drop off as rapidly as in the planar case, Equation IV.21 which drops off as $t^{-1/2}$.

The elimination of wastes is functionally the same, and a cylindrical organism eventually dies if it relies only upon diffusion of wastes into an infinite medium.

THREE DIMENSIONAL DIFFUSION OF NUTRIENT. Microscopic organisms frequently rely upon three dimensional diffusion over short distances to obtain nutrients (e.g. phytoplankton in the ocean). The idealized problem of diffusion of nutrient to a spherical organism is then of particular interest.

In spherical coordinates with spherical symmetry the diffusion equation IV.16 becomes

$$\partial^2 g/\partial r^2 + 2r^{-1}\, \partial g/\partial r - D^{-1}\, \partial g/\partial t = 0, \qquad (IV.28)$$

where r is the spherical distance measured from the center of the organism. The solution to Equation IV.28 subject to the boundary condition IV.11 for $r = a$ and the initial condition IV.12 is

$$g = g_0 - \frac{ba^2 g_0}{(1 + ba)r}[\text{erfc}(X) - \exp(Y^2 - X^2)\, \text{erfc}(Y)],$$
$$(IV.29)$$

where

$$X = (r - a)/2\sqrt{Dt}, \qquad Y = (r - a)/2\sqrt{Dt} +$$
$$(1 + ba)\sqrt{Dt}/a, \qquad (IV.30)$$

and erfc(X) is the error function complement of X.

The rate that nutrients are ingested per unit area, R, Equations IV.10 and IV.29 for $r = a$ is given by

$$R = [kg_0/(1 + ba)]\,[1 + ba\, e^{\tau}\, \text{erfc}(\sqrt{\tau})], \qquad (IV.31)$$

where the dimensionless time parameter $\tau$ is given by

$$\tau = (1 + ba)^2\, D/a^2 t. \qquad (IV.32)$$

The final steady state rate of ingestion per unit area for $t \to \infty$ becomes

$$R_\infty = kg_0/(1 + ba), \qquad (IV.33)$$

where $b = k/D$. If the requirement for nutrient $R_x$ is greater than $R_0 = kg_0$, the initial influx of nutrient, the organism perishes immediately. If $R_\infty < R_x < R_0$, then the organism eventually dies, as was the case for one and two dimensional diffusion. If, however, $R_x$ is sufficiently small so that $R_x < R_\infty$, the spherical organism can survive indefinitely in an infinite medium supplied by diffusion only, as indicated in Figure 7. A similar situation results for the elimination of wastes.

While spherical geometry is the most favorable for mass transport by diffusion alone, it is still generally a slow process. Diffusion as a mechanism of mass transport appears to be significant primarily in the microscopic domain.

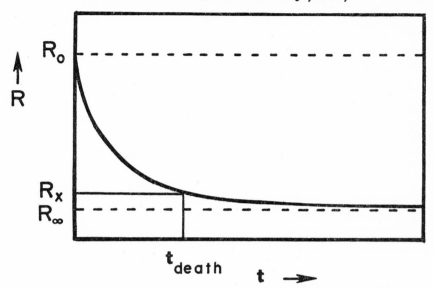

Figure 7. A diagram indicating the rate at which a sessile spherical organism can ingest food (or eliminate wastes) R from an infinite medium if diffusion processes only are present as a function of time t, Equation IV.31. If $R_x$, the minimum time rate for survival, is greater than the initial rate $R_0$ then life will cease immediately. If $R < R_x < R_0$, the organism will eventually die at the time $t_{death}$ indicated. If $R_x < R$, then the organism can live indefinitely.

## Convection and the Ability of an Environment to Support Life

While life can exist in principle if only diffusion processes and steady state sources of nutrient are present, the rate of mass transport is extremely small compared with gross hydrodynamic flow or convection. The type of convection of interest here is that which is effective in transporting nutrients to an organism or wastes from an organism. A sessile organism which is firmly anchored to the earth can enjoy the full transport potential of convective flow of air or water past it. A free-floating, nonmotile organism, on the other hand, would experience no mass transport relative to itself in nonturbulent or laminar flow of air or water. Such free-floating organisms would find nutrients transported to them by convection only if turbulence were present to cause mixing. Effective convection can therefore occur at solid fluid interfaces or in turbulent flow.

EXAMPLES OF CONVECTIVE MASS TRANSPORT. Since the potential for life is considerably enhanced where convective mass transport occurs, it is of interest to examine environments where such transport occurs.

Wind blowing over the solid earth provides $CO_2$ to growing plants. Plants do not generally have to rely upon diffusion of gases (except possibly inside the plant), since even a slight motion of the air is sufficient to replenish the supply of $CO_2$ and to remove the excess $O_2$.

In the ocean far from shore, life is not anchored to the solid earth, so that life in the ocean depends upon turbulent flow. The turbulence and mixing is provided by wind-driven waves. The spray from whitecaps provides a further mixing between the water and the atmosphere. Autotrophs near the surface are, thereby, kept supplied with adequate concentrations of $CO_2$ and $O_2$. The small size of individual ocean autotrophs as compared with autotrophs growing on the land can be accounted for in terms of the difference in the transport mechanisms available to each. On land large autotrophs which are anchored to the earth are readily supplied with large quantities of gaseous nutrients by wind or by gaseous diffusion (which occurs at a rate about $10^4$ times greater than diffusion in water). In the ocean free-floating autotrophs must rely upon a much smaller rate of convective mass transport. They have to be small in order to become intimately involved with the wave turbulence near the surface. Ocean autotrophs must depend upon diffusion for mass transport over short distances. The rate that they absorb nutrients per unit mass and, thus, per unit volume is proportional to

$$3R/a \,, \tag{IV.34}$$

where R is the rate of ingestion per unit area and a is the radius of the individual autotroph. Since R is small by diffusion alone, a must be small. In other words the ocean autotrophs must be small in order to have a sufficiently large specific surface on which to absorb nutrients by diffusion.

Ocean currents can provide convective mass transport. The

abundant life associated with the upwelling of the Humbolt Current off the coast of South America is an example of the life-supporting capacity of such turbulent ocean currents.

In shallow ocean water where the solid-liquid, gas-liquid and possibly gas-solid interfaces all come into play the chance for mass transport is very great, and life can thrive in such regions. The convective mass transport provided by the water in such shallow areas can occur due to tides, waves or surf. Life on the earth may have first evolved in such favorable environments. Sea coasts and coral reefs today present some of the most complex and most evolutionarily advanced ecosystems in the world.

Rain falling on land is responsible for mass transport in many situations. Rain soaks into the ground dissolving nutrients which are carried by capillary action to the roots of land plants. Flow by capillary action, while similar to diffusion in some ways, is much more rapid than ordinary diffusion and can be likened to convective flow. Wherever rain water flows it carries with it the capacity for the mass transport of nutrients by convection. Streams, rivers and small ponds replenished after each rain are all good places to expect life. Within a river the more favorable places for life should be in falls, rapids, eddies and whirlpools where the opportunity for effective gross hydrodynamic mass transport is the greatest.

The flow of water through the earth is responsible for percolating processes which deposit ores. As explained in the previous Chapter, such growing crystalline deposits may be regarded as simple forms of ecophysical life.

The earth's atmosphere is filled with all sorts of small and microscopic forms of life, such as eggs, spores, pollen, bacteria, seeds and protozoa. The role of dust, wind and rain in this environment is not clear. It is also not evident that the atmosphere can be treated as an independent ecosystem.

Even when the wind does not blow, green plants utilize convection for mass transport. Their green leaves absorb sunlight and become hotter than the surrounding air. The air immediately next to the leaf then becomes hot and rises, thus providing mass transport by convection due to heating.

The transport of solid materials by rivers, surf, tides and wind may also play a role in the mass transport processes necessary for life. For example, alluvial deposits are generally more favorable for life than mountain granite.

CONVECTIVE MASS TRANSPORT ON A PLANETARY SCALE. Essentially all of the convective mass transport on the earth can be attributed to atmospheric circulation. The prevailing westerlies and trade winds provide a more-or-less stable pattern of flow, while storms acted on by the Coriolis force provide giant pinwheel patterns which move over the earth's surface. From radioactive fallout from nuclear explosions it has been ascertained that gaseous molecules released anywhere will become more-or-less evenly mixed within a hemisphere after about three or four months, while the rate of mixing between the northern and southern hemispheres is much slower, taking two or more years to produce an even distribution.

The greenhouse effect is apparently the driving force behind the circulation of the atmosphere. Sunlight penetrates the atmosphere and is absorbed on the surface of the earth. Thermal radiation cannot penetrate the atmosphere as readily as the sunlight, so that heat accumulates next to the earth's surface. Equilibrium is eventually established by some of the thermal energy trapped next to the earth's surface being carried aloft by gross hydrodynamical flow of the atmosphere. The heat transported aloft is then radiated off into space at a higher altitude where there is less atmosphere to impede its transport into outer space.

Water on the surface of the earth absorbs solar energy and is evaporated. The water vapor together with its energy is then carried aloft where the energy is released and radiated off into space, the water condensing out and falling back to earth as rain. The mass transport produced by rainfall is, therefore, also a product of the general greenhouse effect.

An accurate theoretical estimate of the magnitude of the greenhouse effect on the earth and on the other planets of the solar system is extremely involved, and there appears to be little agreement among various estimates found in the literature. A rough measure of the mass transport produced by the greenhouse effect will be taken here as the mean rate of mass flow per unit area, $\rho v$, where v is some mean hydrodynamic velocity of the atmosphere near the surface of the planet and $\rho$ is the mass density. It may be assumed that a certain fraction of the energy trapped next to the surface of the planet appears as the kinetic energy of hydrodynamic flow,

$$K = \rho v^2/2 . \qquad (IV.35)$$

If the kinetic energy of hydrodynamic flow can be estimated the rate of mass flow per unit area becomes

$$\rho v = (2\rho K)^{\frac{1}{2}} . \qquad (IV.36)$$

The rate that energy is carried aloft per unit area by convection B should be proportional to the amount of energy trapped near the surface of the planet and should, therefore, be proportional to the kinetic energy of hydrodynamic flow near the planet's surface; or $B \propto K$. Using the earth as a standard, a dimensionless measure of the amount of effective convective mass transport that exists on a planet is then given by

$$J = (\rho B / \rho_e B_e)^{\frac{1}{2}} , \qquad (IV.37)$$

where $\rho_e$ and $B_e$ are values estimated for the earth. Estimated values of J for the various planets in the solar system are included in Table I.

### Motility to Effect Requisite Mass Transport

Under the heading of motility two types of phenomena are considered: (1) self-propelled movement of an organism through a medium or across a surface to effect ingestion of nutrients or to eliminate wastes, and (2) the forced convection of material by the action of an organism to effect ingestion or elimination.

A true autotroph cannot increase its energy consumption, i.e. amount of sunlight absorbed, by motility, since every local area about the autotroph receives precisely the same amount of sunlight. The motion of leaves to face the sun represents such a slight motility that it need not be of concern here. Other requirements for life, such as essential salts or water, have not led to motile adaptations in autotrophs. It appears that in regions on the earth's surface where autotrophs cannot survive motile autotrophs would be equally incapable of surviving. Motility is a feature limited primarily to heterotrophs.

The slightest convection of material by a stationary organism quickly overcomes any limitation of mass transport due to diffusion only. Precisely the same result of convection in the environment is achieved if the organism moves while the medium remains stationary. Consequently, due to the advantage of motility, motility among heterotrophs should be expected to be a common evolutionary development. If a motile organism requires nutrient at a rate of S grams per second, then it will have to move through a stationary medium (neglecting the negligible effect of diffusion) at a minimum velocity v in centimeters per second such that, if the effective cross sectional area for ingestion is $\sigma$ in centimeters squared and the density of nutrient is g in grams per centimeter cubed, then

$$v = S/\sigma\, g\,. \tag{IV.38}$$

The relationship also holds in the two dimensional case for a predator seeking prey on a two dimensional surface, where g is the surface density of prey and $\sigma$ is a linear capture distance.

In addition to motion of an organism through a medium, organism can produce mass transport by the forced convection of material. This type of adaptation includes the motion of barnacles to produce a flow of sea water and prey through their ingestion organs, peristalsis of gullets and intestines, waving motion of cilia, the one-hole bellows action of lungs and sea squirts, pumping action of hearts, contractile motion of bladders and vacuoles, the two-hole bellows action of gills, etc. It may again be assumed that the advantage gained by such forced convection should tend to favor the evolution of heterotrophs that can produce such flow.

The adaptation of motility greatly expands the number and types of environments available to life, especially sophisticated forms of life such as man and machines.

### Nonmotile Mass Transport Produced by Plants

There is important mass transport produced by plants which does not require moving organs or motility. Plants evaporate water from their leaves. The process of osmosis then raises large quantities of water and mineral nutrients from the soil up into the plant. Tall trees produce an impressive flow of water and minerals to great heights by large negative pressures which are maintained by some processes yet unknown. The growth of plants, particularly roots, can displace dirt and large rocks by a sort of hydraulic action. While this may not constitute a very large mass transport, it is important in soil conditioning.

## *Extraterrestrial Environments*

One of the more dramatic reasons for investigating the environmental conditions necessary for life is to estimate the likelihood of finding life in the universe apart from the earth. Research in exobiology to date has been largely limited to the consideration of carbon-based life (e.g. Lederberg, 1960; Sagan, 1961; Shneour and Ottesen, 1966; Saunders, 1969; Dobzhansky, 1972). Investigations concerning the origin of life have been largely limited to the consideration of the origin of certain organic compounds such as amino acids and polypeptides (Fox, 1971; Schwartz, 1971). The search for extraterrestrial life using rockets, as currently planned, is largely limited to the search for certain carbon compounds (for example, optically active compounds [Stryer, 1966]).

While this preoccupation with carbon chemistry is perhaps understandable in terms of carbon-based life as it is ordinarily known on the earth, it does not envision life in a context sufficiently broad to yield a proper estimate of what may actually be found on other planets. Present knowledge concerning extraterrestrial environments is largely limited to rough estimates of their thermodynamic properties. Consequently, the present ecophysical definition of life and the study of thermodynamic and other physical properties of environments that can support life are particularly appropriate for the search for extraterrestrial life.

Likelihood of Life Evolving in the Solar System. Table I presents a number of characteristics of environments in the solar system that may be used to estimate the likelihood of life evolving in these environments. The various columns in the table show:

TABLE I. SOME CHARACTERISTICS OF BODIES IN THE SOLAR SYSTEM GIVING AN INDICATION OF THE LIKELIHOOD OF LIFE*

| body | $r$ cm $\times 10^{13}$ | $A$ | $T$ °K | $T_g$ °K | $P$ $\frac{gm}{sec^3 \,°K}$ | presence of gas | liquid | solid | temporal changes | $\rho$ $\frac{gm}{cm^3}$ | $Q$ $\frac{erg}{cm^2sec}$ $\times 10^3$ | $J$ |
|---|---|---|---|---|---|---|---|---|---|---|---|---|
| Sun | 0.00696 | — | $10^6$ | 5800 | $1.10 \times 10^7$ | x | 0 | 0 | violent | $10^{-6}$ | $6.44 \times 10^7$ | 106 |
| Mercury | 0.579 | .056 | 620† | 626† | 6290 | 0 | 0 | x | none | 0 | 2200 | 0 |
| Venus | 1.081 | .76 | 650 | 230 | 435 | x | 0 | 0 ? | cloud | 0.02 | 160 | 24 |
| Earth | 1.495 | .36 | 290 | 250 | 1450 | x | x | x | cloud seasonal | 0.0012 | 223 | 1 |
| Moon | 1.495 | .067 | 388† | 388† | 1560 | 0 | 0 | x | rare | 0 | 325 | 0 |
| Mars | 2.278 | .16 | 230 | 217 | 1050 | x | 0 | x | cloud seasonal | $2 \times 10^{-5}$ | 126 | 0.021 |
| Ceres | 4.186 | .06 | 234† | 234† | 341 | 0 | 0 | x | none | 0 | 41.8 | 0 |
| Jupiter | 7.778 | .73 | 140 | 88 | 48 | x | x | 0 ? | cloud | 0.1 | 3.5 | 7.8 |
| Saturn | 14.26 | .65 | 120 | 70 | 22 | x | x | 0 ? | slight | 0.1 | 1.3 | 4.8 |
| Uranus | 28.68 | .66 | 70 | 49 | 9.0 | x | x | 0 ? | none | 0.1 | 0.32 | 2.4 |
| Neptune | 44.94 | .65 | 50 | 39 | 5.3 | x | x | 0 ? | none | 0.1 | 0.14 | 1.5 |
| Pluto | 59.08 | .14 | 60† | 60† | 6.3 | 0 | 0 | x | none | 0 | 0.19 | 0 |

*See text for explanation of symbols.

†Daytime only.

*r,* the distance from the center of the sun (the radius of the sun in the case of the sun itself).

*A,* the spherical or Bond albedo as estimated from a number of sources (e.g. Kaula, 1968).

*T,* the observed surface temperature as estimated from a number of sources (e.g. Kaula, 1968). The temperature measured for Jupiter, Saturn, Uranus and Neptune probably represent temperatures above their solid surfaces (if, in fact, they have any solid surfaces). The temperature at the bottom of the outer turbulent layer of the sun is roughly estimated to be $10^6$ ° K.

$T_q$, the surface temperature that would be present if there were no atmosphere producing a greenhouse effect, obtained by assuming thermal equilibrium and reradiation as an ideal black body; thus,

$$T_q = [(\mathrm{R}r_e^2/4\,\sigma)(1-A)/r^2]^{\frac{1}{4}} = [1.36 \times 10^{36}(1-A)/r^2]^{\frac{1}{4}},$$
(IV.39)

where all quantities are in cgs units, $r_e$ is the distance of the earth to the sun (the astronomical unit), $R = 1.395 \times 10^6$ erg/cm$^2$ sec is the solar constant, and $\sigma = 5.670 \times 10^{-5}$ erg/cm$^2$ sec (° K)$^4$ is the Stefan-Boltzmann constant. For bodies with no atmospheres—Mercury, the moon, Ceres and Pluto—the equilibrium is assumed to be established immediately, so that the bodies radiate as discs illuminated only on the daylight side, Equation IV.39 being multiplied by $\sqrt{2}$ in these cases. The outer surface of the sun is assumed to have a black body temperature of 5800° K.

*P,* the entropy production per unit area, estimated by Equation IV.3, or

$$P = 1.55 \times 10^{32}(1-A)(T^{-1} - 1/5800)/r^2 . \qquad (IV.40)$$

The entropy production in the surface of the sun per unit area is given by the outward flux of energy per unit area with an inside temperature estimated at $10^{6}$° K and an outside temperature of 5800° K.

*Presence of gas, liquid or solid,* the presence of gas, liquid or solid in the entropy producing environment. Gas and liquid indicate the possibility of transport mechanisms being present. A solid is counted as being present if the solid surface is involved with sunlight or entropy production. It is assumed that solids are necessary for life, since the ecophysical definition of life requires internal states of low entropy.

*Temporal changes,* changes which are actually observed. Life, being a dynamic ongoing process, might be expected to be associated with temporal changes, such as seasonal changes. Any observed temporal changes might be suggestive of the possibility of life.

$\rho$, the density of the atmosphere next to the planetary surface. These estimates are very crude, especially for Jupiter, Saturn, Uranus and Neptune where no solid surface may exist.

$Q$, the energy passing through the environment per unit area per unit time averaged over the surface of the planet as measured by the rate of absorption of sunlight; thus,

$$Q = R \, r_s^2 (1 - A)/4r^2 = 7.79 \times 10^{31} \, (1 - A)/r^2 \, , (IV.41)$$

in cgs units. The factor of 4 is absent in the denominator for the case of the sun. The rate that energy is made available to the environment also helps to indicate the potential for life.

$J$, the dimensionless measure of the atmospheric mixing given by Equation IV.37 where $B_e$ is roughly estimated to be 25 percent of the absorbed solar energy per second per unit area, giving

$$J = 0.42 \sqrt{\rho \, B} \, . \tag{IV.42}$$

It is roughly estimated that about 0.1 percent of the absorbed solar radiation on Mars is carried aloft by convection. On Venus, Jupiter, Saturn, Uranus and Neptune it is assumed that 100 percent of the absorbed radiation is carried aloft by convection. All of the energy transmitted through the outer turbulent layer of the sun is assumed to be carried aloft by convection.

From Table I it is apparent that the outer layer of the sun provides a tremendous entropy production per unit area. It also has a violent rate of convective mixing as indicated by J in the

table. On a large time scale the evolution of any star, and therefore, the sun, is in the direction of decreasing internal entropy.
The only reason for doubting the existence of life on the sun is
the lack of solids. However, the entropy production produces ordered structures such as convection cells and sunspots. The
cooler gas within a sunspot indicates a refrigerating mechanism is present. Inside the surface of the sun, where the tubes
of cooler gas cannot be seen, is the region in which the entropy
production largely takes place. The mechanism giving rise to
the spot, thus remains hidden from view. While it may seem
unlikely, yet there is still a possibility that inside the sun, where
the magnetic fields are much stronger and where the refrigerating process is actually taking place, solids do occur. Consequently, there is a possibility that life may exist on the sun. It
should also be recalled that the sunspots may be several times
the diameter of the earth, so that the volume in which ordering
processes can occur is considerable. A spaceship equipped to
keep itself refrigerated in a 5800 °K plasma could conceivably
explore the surface of the sun and possibly utilize the solar energy to run the refrigerating mechanism. The problem of inventing an appropriate refrigerating mechanism for a 5800° K
plasma is not as difficult as it might at first appear. The plasma
could be handled with magnetic fields and, because of the large
volume available, the fields need not even be particularly strong.
Electric currents running through the plasma in the presence of
magnetic fields could then provide electromagnetic pumping action. Centrifugal separation of hot gases from cold gases could
then conceivably keep the spaceship bathed in a small cloud of
cold gas (which would have to be continually replenished to
remove absorbed radiation).

Mercury is undoubtedly a dead planet due to the complete
lack of transport mechanisms. While the lack of mass transport
mechanisms can be overcome by motile life forms, it seems impossible that life could ever evolve organs for motility from the
very start in an environment devoid of all mass transport mechanisms. It is true that life that evolved elsewhere could conceivably utilize its motility to live on Mercury; the likelihood of
such *sophisticated* life being on Mercury will be considered in

a subsequent section. Since only solids exist on Mercury, mass transport is limited to solid diffusion under the action of sunlight and the solar wind. The slow rate of solid diffusion means that during the life of the solar system, about $4.5 \times 10^9$ years, there has not been sufficient time for Mercury to have attained any significant amount of ordering by the action of sunlight.

Venus probably lacks life on its solid surface, since it is doubtful that any significant amount of sunlight reaches the solid surface. The rate of hydrodynamic mixing or convection is extremely great as indicated by the value 24 for J listed in Table I and the observed temporal cloud changes. The small entropy production per unit area (which may not even be associated with the solid surface) indicates a very small probability for life existing. Life, if it exists on Venus, is probably primitive and probably lives high up in the atmosphere supported on the wind.

Earth has a high rate of entropy production per unit area; its solid surface is illuminated; it has liquids as well as gases present; and it has a moderate amount of hydrodynamic mixing. Considering only the physical features of the environment, it appears that the earth provides the greatest potential for life in the solar system. This means that all ordering processes powered by the sun have a greater potential for occurring on the earth; thus, a search for ores or accumulation of separated compounds should be most successful on the earth. Since life is known to occur on the earth, it might be thought that perhaps the definition of *life* has been chosen here so narrowly that only the earth could conceivably support such *life* in the solar system. While this might be true of many proposed definitions of life, the ecophysical definition adopted here and presented in the previous Chapter is based only upon a few broad thermodynamic properties. The superior quality of the earth's environment as determined from these broad thermodynamic principles is, therefore, quite significant and surprising. It also indicates the disappointing likelihood that no advanced life exists anywhere else in the solar system no matter how different its structure or chemistry.

The moon, like Mercury, lacks any mass transport mechanisms to give rise to life. The direct observations of the Apollo flights to the moon have verified this expected absence of life on the moon. (For further speculation see Gilvarry, 1965.)

Mars has about the same entropy production per unit area as the earth, being only about 25 percent less. The observation of seasonal changes in color and the seasonal changes in the polar ice (presumably solid $CO_2$) appear to indicate that at least eco-physical life in its lowest form of simple crystal growth does occur. It might be argued that organisms should evolve to utilize as much sunlight as possible, which would simply imply that with time the surface of a planet with life should tend to absorb more sunlight and to develop a lower albedo. Mars has a low albedo which might be interpreted as indicating a greater likelihood for the existence of life. Unfortunately, the bodies in the solar system with no atmosphere tend to all have very low albedos and, of course, no life. The solid surface of Mars is illuminated by sunlight, a condition favorable for life. There is, however, very little chance of any advanced life existing on Mars due to the negligible amount of convection or gross hydrodynamic mixing that can occur, as indicated by the small value of 0.021 for J from Table I (which is probably an overestimate). Observations from the earth and Mariner 9 of high velocity dust storms on Mars might be interpreted as indicating a large rate of mass transport of dust, but the persistence of meteoric impact features on Mars indicates that the actual rate of mass transport of dust must be quite small. Since liquids are apparently absent, life on Mars would have to be based primarily upon solid-gas transitions. Life on Mars is undoubtedly very meager and of a crystalline form (for further speculation see Jackson, 1965).

The asteroids such as Ceres, lacking atmosphere and mass transport mechanisms could not evolve life.

The entropy production per unit area on Jupiter is small, being only about one-thirtieth that of the earth. Even if it is assumed (as a consequence of some recent temperature measurements) that Jupiter is, in fact, a radiating body like a protostar, the entropy production remains roughly the same, since the amount of energy that would be radiated is comparable to the incident solar radiation absorbed. Sunlight never penetrates to the solid surface (if it has a solid surface); however, the permanence of the great orange spot indicates that floating solid material might be irradiated by sunlight. Jupiter experiences a large amount of mass transport as indicated by gross hydrodynamic mixing. The frequent temporal changes, the variegated appearance, and the time-varying radio signals are all suggestive of the presence of life. If life exists it probably occurs as simple crystalline life on floating particles supported by the wind.

The likelihood of life on Saturn is considerably less than on Jupiter. The likelihood of life on Uranus, Neptune and Pluto appears to be vanishingly small.

LIKELIHOOD OF LIFE ON COMETS. Life on comets appears to be very unlikely despite the presence of organic compounds and probably water. Comets receive significant amounts of sunlight only briefly as they swing close to the sun. For most of their orbit they are too far out for sunlight to achieve any significant ordering. In addition, the gravitational field of a comet is so weak that gaseous materials, necessary for mass transport, are boiled off into space as the comet swings close to the sun.

It was once thought that comets were from beyond the solar system and that they might introduce life from beyond the solar system. It is now generally believed that comets are actually a part of our solar system. The possibility of life being introduced from without our solar system is considered below.

LIKELIHOOD OF ORDINARY LIFE FROM BEYOND THE SOLAR SYSTEM. It is conceivable that primitive life was somehow transported from beyond the solar system and became established in particularly favorable environments within the solar system (cosmogenesis by panspermea). Once such initial *seed* or *germ* became established, evolution could then have produced advanced life forms. This possibility is of little interest, however, since the process of evolution produces life forms dictated by the needs of the ecological niche independent of the nature of the original seed. It makes little difference whether the initial seed is assumed to have an origin from beyond the solar system or whether the initial seed is assumed to have generated *spontaneously* within the solar system; the end results are the same.

If it is assumed that more advanced life immigrated to the solar system, it is of some interest to ask what part of the solar system provides suitable environments and, thus, where such advanced life might be found. It might be argued that such life would seek out environments similar to those in which it originally evolved. But such life could have conceivably also evolved in the local environment (i.e. in the solar system), the conditions being comparable. Consequently, the likelihood of such life evolving locally becomes about the same as the likelihood of such life evolving in a similar environment beyond the solar system and being transported into the solar system. The region of the universe within any reasonable radius of the solar system may be assumed to have roughly the same age as the solar system, 4.5 billion years (the age of the universe being no more than about 10 billion years). It, therefore, seems unlikely that life suitable for a particular environment could have evolved more rapidly elsewhere and then have had time to be transported to the solar system before advanced local fauna had already evolved. The question of whether or not primitive or ordinary life forms came from beyond the solar system is largely a matter of indifference, since it is equivalent to the question of the likelihood of life arising spontaneously within the solar system.

LIKELIHOOD OF SOPHISTICATED LIFE FROM BEYOND THE SOLAR SYSTEM. Life which can survive in a wide range of environments will be regarded as *sophisticated*. For example, man together with machines, spaceships and elaborate technology represents sophisticated life. Since sophisticated life is the only type of life that could conceivably have a technology sufficiently advanced to transport itself across interstellar distances, it is apparently the only type of life from beyond the solar system that needs to be considered seriously.

Sophisticated life can have a very rapid evolutionary rate, as indicated by the rapid evolution of machines (see Chapter IX), so that sophisticated life from beyond the solar system could conceivably be more advanced than the life which has evolved within the solar system.

If such life obtained its high utility energy from contained nuclear reactions (such as fission or fusion), it seems unlikely that such life would ever enter the solar system (except possibly to obtain matter). If, however, such contained nuclear reactions resulted in a poisoning of their environment with radio-active wastes, such life might prefer a nuclear reactor such as the sun. The sun obtains its energy at a low temperature as compared with the temperature needed in thermonuclear devices for the direct conversion of deuterium to helium. The low internal temperature of the sun, in addition to its cool envelope of only 5800° K, provides a safe source of steady energy which need not radioactively poison the environment.

If sophisticated life were to choose to utilize solar energy, it may be assumed that it would seek the optimum environment for the utilization of sunlight. A search for sophisticated life in the solar system is then a search for the optimum environment.

THE OPTIMUM ENVIRONMENT IN THE SOLAR SYSTEM FOR SO-PHISTICATED LIFE. The optimum environment appears to be in space revolving around the sun in free fall. The surface of a planet presents a number of awkward features which seriously detract from its desirability.

A planet which rotates has periods of light and dark. During the dark periods no energy is available, so that living processes must either cease during the night or else energy must be stored during the day to be utilized at night. In either case living processes are seriously handicapped. At the very best the high utility solar energy available is halved in rate.

On the surface of a planet with an atmosphere the sink temperature for the dissipation of low utility energy is much higher than the temperature of deep space which may be about 3.5° K, so that the efficiency of utilization of solar energy is decreased. The energy available on most planets with an atmosphere is also considerably reduced by reflection from clouds and absorption in the atmosphere.

In free fall the transport of matter necessary for the ordering processes of life can take place along geodesics without the expenditure of any energy. On a planet gravitational forces give rise to frictional forces that must be overcome to transport materials. Man and machines, for example, consume a large fraction of their total energy expenditure to transport men, machines and goods across the surface of the earth.

The mass transport processes on the surface of a planet, being restricted to the surface of a planet, are essentially two dimensional, whereas in space, mass transport processes can become three dimensional. The ordering of materials in a three dimensional array can occur over much shorter distances, at much faster rates, and with much less effort than is possible for two dimensional ordering processes.

The time rate that materials can be transported in a vacuum can be large; the frictional drag of an atmosphere would not interfere.

The absence of an atmosphere permits an easy control over the temperature that can be readily obtained. By shielding a body from the direct rays of the sun it is possible to obtain cryogenic temperatures down to perhaps 10° K. And by using a parabolic mirror to focus the sun's rays, it is possible to obtain temperatures approaching the temperature of the sun's surface of 5800° K (reradiation limits the upper temperature attainable to about 4500°K). Elaborate and costly systems must be constructed to achieve similar temperatures on the surface of a planet, and such temperatures can be obtained only in very small volumes and only with a continuous expenditure of energy.

In free fall structures and buildings need not have the strength to overcome the effects of gravity. The only limitations in space on the strength of structure is prescribed by the impulses imparted to the structures by the users. Structures need not be severely limited in size as is the case, for example, on the earth.

A vacuum, being readily available in outer space, permits a wide variety of technologies presently expensive or impossible on the surface of a planet with an atmosphere.

Direct line-of-sight communication, which is impossible on the surface of a planet, is readily available in space.

There are a few apparent disadvantages to living in space. An environment requiring volatile liquids and gases under pressure needs a containing envelope of some strength. No such envelopes are needed on the surface of a planet which can contain an atmosphere at the required pressure by gravitational attraction. However, it might be feasible to contain gases in space by using a plasma envelope which is confined by a magnetic field established over a large region of space. The magnetic fields could be maintained by superconducting magnets with no expenditure of energy, since cryogenic temperatures are available in space. It should be noted that truly sophisticated life, machine life, has little or no need for such gaseous environments.

All of the matter that may be required in space must be taken from a body in the solar system with a great deal of effort and a large expenditure of energy. However, once the matter has been placed in the appropriate orbit about the sun, it will be available forever after without any additional expenditure of energy. The matter may, thus, be accumulated slowly over a period of time, and no large drain upon the reserves of the sophisticated life need be assumed. For example, the earth is now encircled by approximately two thousand artificial satellites accumulated over the last sixteen years.

Cosmic rays, as well as soft x rays from the sun, might produce disordering effects upon sophisticated life. However, magnetic fields established over large volumes of space could effectively shield life from energetic charged particles, and low density gas contained in a large region could effectively stop soft x rays. The fields could be maintained without any energy expenditure by using superconducting magnets. None of the disadvantages to living in space appear to be insurmountable for sophisticated life. The advantages to living in space, as opposed to living on the surface of a planet, appear to outweigh the disadvantages.

The optimum distance from the sun for the utilization of sunlight is probably inside the orbit of Mercury (Wesley, 1966). Since there appears to be no evidence of any objects inside the orbit of Mercury which might obstruct the flow of sunlight, there appears to be no evidence for the presence of sophisticated life superior to that on the earth today, specifically life from beyond the solar system.

ENVIRONMENTS BEYOND THE SOLAR SYSTEM. It is reasonable to assume that other stars also have planets and that these planets will have a variety of potentials for supporting life such as found in the solar system. But there are also some extremely different types of environments beyond the solar system that may possess life. Toward the end of stellar evolution a star becomes a white dwarf. A white dwarf contains *crushed* matter which takes on the character of a solid which has a state of low enough entropy to admit the possibility of ecophysical life, according to the definition presented in the previous Chapter. The energy flux from such a white dwarf is still sufficient to support life. A neutron star, such as the pulsar in the Crab Nebula, also has the possibility of a solid-like internal structure and also possesses the possibility of life. Quasars are such strange objects that the possibility of associating life with quasars cannot be entirely ruled out.

While it is easy enough to discount such exotic environments as potential environments for life, it may not be wise to do so. A search for extraterrestrial intelligence, for example, might profit by a close examination of the radiation from white dwarfs, neutron stars and quasars. The ecophysical concept of life is not bound by the nature of ordinary carbon-based life as found on the surface of the earth. Once the general principle of physics have been employed, they are applicable to all natural phenomena, including the earth, white dwarfs, neutron stars and quasars.

# CHAPTER V

# SOURCES OF ENERGY

As EXPLAINED IN THE PREVIOUS CHAPTERS, life can survive only as long as there is a source of high utility energy that can be dissipated to low utility energy. A source of high utility energy that can be dissipated to low utility energy then provides a potential for life. The examination of possible energy sources then indicates where and how life might exist, especially men, machines and possible alien forms of sophisticated life.

## Value of Energy

The basic *value* or *good* as far as life is concerned may be taken as the creation and preservation of a state of thermodynamic order. The value of energy may then be measured in terms of the amount of order it can create and for how long a time. The degree of order that can be created and the length of time it can be maintained is not only proportional to the amount of energy available but it is also dependent upon the efficiency with which the energy is utilized in an engine or in an organism, as well as the cost to extract the energy from its source, the cost to store energy until needed, and the cost to transmit the energy to the point where it can be used. The yardstick or measure of any general value is usually in terms of the most valuable objects or commodities generally available. Work is energy in its most valuable form, since work can create the greatest amount of thermodynamic order as indicated in Chapter I, Equation I.22. Therefore, the measure of the value of energy may be taken as the amount of work that in principle could be made available at the site where it is to be used per unit of the energy dissipated. The value of an energy source is, thus, first of all, given by the efficiency with which it can in principle be converted to work.

A source of thermal energy can be converted to work with an efficiency no greater than that of a Carnot engine, Equation I.14. Actual irreversible heat engines operate at much lower efficiencies (*cf.* Equation II.16 for $T_1 = T_1{}^*$, Equation II.13). For an environment with an atmosphere the temperature of the atmosphere determines the ambient temperature which is then generally the temperature of the cold reservoir available. On the surface of the earth the conversion of thermal energy to useful work is accomplished with a much lower efficiency than could be achieved in outer space where a cold reservoir close to $10°$ K is readily available. Many large reservoirs of thermal energy are available on the earth at temperatures so little different from the atmospheric temperature that little or no useful work can ever be extracted. Such energy sources are of little value.

Since the living process is an ongoing process, the value of an energy source also implies its time rate of utilization. If profit is measured in units of work, the ultimate measure of the value of energy source should be the time rate of profit that could be realized in principle if the energy source were utilized. Considering the cost to procure the energy and the price of the work obtained, the time rate of profit is given by Equation II.9 (or Equation II.7).

The utilization of an energy source by man sometimes depends upon the initial cost of development or the initial cost of construction. The present-day limited utilization of nuclear energy and solar energy, which both show promising futures, arises from high development and construction costs. While development costs may be difficult to estimate, the construction cost of a power plant of known design and performance is not difficult to estimate. If the funding of the initial construction cost of a power plant is achieved by borrowing money (and/or selling stocks), then the funds obtained depend upon the rate of interest (the expectation of profit or return on investment) and the rate of retirement of the indebtedness. In particular, if X is the amount owed (and/or shares outstanding) at any time, then the rate of profit is given by

$$\dot{R} = A\dot{W} - B\dot{Q} - (i + k)X , \qquad (V.1)$$

where $\dot{W}$ is the rate of work delivered at the price A per unit of work, $\dot{Q}$ is the rate of consumption of energy at the cost B per unit of energy, i is the rate of interest (or return on investment), and k is the rate of retirement of the indebtedness X.

If the debt is never retired, $k = 0$, and $R \geqslant 0$, so that

$$X \leqslant (A\dot{W} - B\dot{Q})/i . \qquad (V.2)$$

For example, nuclear or solar energy is essentially free, $B \approx 0$. If the price of electricity is to be 5 mills per kilowatt hour (the value of work and electrical energy being almost identical) and the interest rate is 2.5 percent per year (inflation being discounted here), then the initial cost of such power plants could be as high as $1700 per kilowatt of power capacity. If, however, the indebtedness is being fractionally reduced at the rate k, the amount owed at any time becomes

$$X = X_o\, e^{-kt} . \qquad (V.3)$$

The particular way that the debt is paid off is of no great consequence to the present discussion; Equation V.3 will suffice.

The expectation of gain $R_e$ as viewed at the present time $t = 0$ may be estimated by

$$R_e = \int_0^\infty \dot{R}(t)\, e^{-jt}\, dt , \qquad (V.4)$$

where the exponential factor is included to take into account the probability of loss due to catastrophes or unforeseen circumstances such as war, broken contracts, theft, sabotage, unexpected loss of a market, vandalism, hurricane, unexpected equipment failure, etc. The constant j measures the fractional expectation of loss from all such causes per unit time. The reciprocal $\tau = 1/j$ might be called the *economic expectation half-life*. It may be noted that the expectation of gain, Equation V.4, always remains finite (as it should), since $\dot{R}(t)$ remains bounded and the integrand goes exponentially to zero. The factor $e^{-jt}$ may also be viewed as valuing future profits less than present profit, the further in the future the less the value. For example, a man might wish to evaluate all profits beyond the time of his own death as negligible; then $e^{-jt}$ may be interpreted as measuring his expectation of survival at time t. Substituting Equation V.3 into V.1 and the result into V.4 and integrating yields

$$R_e = (A\dot{W} - B\dot{Q})/j - (i + k)X_o/(j + k) . \qquad (V.5)$$

In order to invest in a project the expectation of gain must be greater or equal to zero, which yields the condition for the initial cost

$$X_o \leq (j + k)(A\dot{W} - B\dot{Q})/j(i + k) . \qquad (V.6)$$

The situation that allows for the largest initial cost is for the debt to be paid off at once at $t = 0$; or letting $k \to \infty$,

$$X_o \leq (A\dot{W} - B\dot{Q})/j . \qquad (V.7)$$

Ordinarily it may be expected that j will be roughly the same as the interest rate i (exclusive of inflation), and the same result is obtained as for the simple case above, Equation V.2.

Human cultures evolve according to the same rules as life in general (see Chapter VIII). The culture that commands the greatest utilization of energy can create and maintain the greatest order and should eventually replace other cultures. The two largest sources of energy to be exploited are nuclear energy and solar energy. Nuclear energy requires large initial outlays for development while solar energy requires large initial investments. In either case, a long economic expectation half-life (or small j) is required. The culture that attains the greatest stability can achieve the longest economic half-life (or smallest j). Such a culture will be the first to fully exploit these largest energy sources and will, consequently, tend to replace other cultures. If man's short life span limits the economic expectation half-life, then it is reasonable to expect that machines will eventually have to plan for the future (see Chapter IX) in order to fully utilize these energy sources.

The September 1971 issue of *Scientific American* has eleven interesting articles devoted to energy and power, particularly in the biosphere and in relationship to man's values.

### Solar Energy

The main source of energy available on the earth are nuclear energy and solar energy (e.g. Daniels, 1964), energy from all other sources being negligible by comparison. Most of the usual sources of energy on the earth are derived directly or indirectly from sunlight, such as the energy in wood, wind, river currents, fossil fuels and water power. While nuclear energy may appear more attractive at the moment, in the long run solar energy will probably become the more important. The sun is, after all, a nuclear reactor that is safe, pollution free, and much vaster than any earth-bound reactor could ever be. Even if all of the uranium and other fissile fuels were used up and all of the fusion materials such as dueterium burned and finally if the whole earth were converted to iron to extract the last bit of nuclear energy, it would still only provide less than a millionth of the energy to be yielded from the sun.

The flux of solar energy at the distance of the earth from the sun is $1.40 \times 10^6$ ergs/cm$^2$ sec ($2.00$ gm cal/cm$^2$ min), the *solar constant*. The whole earth intercepts $1.78 \times 10^{24}$ ergs/sec of this energy. This amounts to an average of $3.50 \times 10^5$ ergs/cm$^2$ sec over the earth incident above the atmosphere. The amount per unit area varies with the cos $\Phi$ where $\Phi$ is the angle measured from the ecliptic (the plane of the earth's orbit about the sun). The angle $\Phi$ as measured with respect to the earth's equator has a yearly variation depending upon the tilt of the earth's axis with respect to the ecliptic. About 35 percent of incident solar energy is reflected back into space as visible light (the effective or Bond albedo), the reflectivity varying with $\Phi$, clouds, haze, snow, etc. About 12 percent is absorbed in the earth's atmosphere on the average and is converted to thermal energy. The temporal and geographic variability of absorption in the atmosphere due to clouds, smoke, haze and water vapor is quite large. The light energy incident per unit area on the ground, the *isolation*, is about $1.86 \times 10^5$ ergs/cm$^2$ sec averaged over the earth, the total energy flux at the ground being $9.44 \times 10^{23}$ ergs/sec. This large quantity of energy may be appreciated by noting that a world population of 4 billion ($10^9$) provides 23,600 kilowatts of sunlight per person, the per capita consumption in the United States being about 11 kilowatts in 1970. A utilization of $5 \times 10^{-4}$ of this potentially available energy would fulfill the world's needs at the United States 1970 level.

The only commercial use of direct sunlight at the present is limited to electrical energy produced from silicon-germanium thermocouples or solar cells that can convert up to 10 percent of the sun's incident energy. Such solar cells with batteries are used to supply energy for the amplification of telephone conversations in remote areas that lack a regular supply of electricity. Such cells are also used in spacecraft and artificial satellites to power radio and television receivers and transmitters. Communication satellites furnish an outstanding example. Because of the high initial cost and the low efficiency such photovoltaic cells do not present a practical method for the development of large quantities of power.

The primary drawback in the direct utilization of solar energy on the surface of the earth is its intermittent character, on during the day and off at night, less during cloudy days and more during sunny days. If solar energy is converted directly to electrical or mechanical energy, during a sunny day over half of the power must be expended to store energy for at least twelve hours. This might be done by raising water from a low reservoir to a high reservoir. Since the mechanical power developed during the day is over twice that which will be finally delivered, the capital costs are more than doubled compared with a temporally firm source of power. In addition, money must be invested for storage of energy for the night and cloudy days and for the conversion of the stored energy for delivery. Capital costs for direct utilization of solar energy to obtain electrical power is possibly four to ten times as expensive as a comparable firm source of power.

It has been proposed to collect solar energy in heat traps where visible light is admitted freely, but once it has been absorbed, the resulting thermal energy cannot escape (an ex-

aggerated greenhouse effect). Thermal energy at a temperature
of perhaps 540° C could then be stored in molten salts to be
extracted as needed. If the heat can be converted to electrical
energy with an efficiency of 40 percent, then an overall efficiency
of 25 to 35 percent might be achieved. The initial costs involving a steam turbine, heat storage, pumps and collectors might
be about three to six times the cost of a power plant of the same
capacity run on coal. Since the solar energy is free, all costs
are initial and maintenance costs and, therefore, such a power
plant might be economically feasible.

Apart from problems of storage, the most efficient direct conversion of solar energy to electrical energy appears to be with
the use of mirrors to concentrate the sun's rays. With ideal
mirrors the surface temperature of the sun, 5800° K, could in

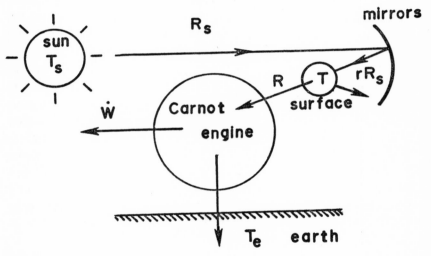

Figure 8.

principle be duplicated; due to reradiation the maximum obtained has been about 4500° K. At such high temperatures ionized plasmas can be obtained and the energy extracted by mag-
netohydrodynamic devices with theoretical efficiencies close to
the ideal Carnot engine. For the ideal case the energy flux incident upon an absorbing surface is given by

$$R_s = \sigma \, T_s^4 , \qquad (V.8)$$

where $R_s$ equals the flux from the sun's surface, $\sigma$ is the Stefan-Boltzman constant, $5.669 \times 10^{-5}$ erg/cm$^2$ sec$^\circ$ K$^4$, and where $T_s$ is the effective temperature of the sun's surface, 5800$^\circ$ K. This may be seen to be correct, since, if the surface were allowed to heat up, it would eventually come into equilibrium at the sun's temperature, as much energy being reradiated as absorbed. For real mirrors with a reflectivity r and the irradiated surface with an absorptivity a at the temperature $T < T_s$, the net rate of absorbing energy per unit area becomes

$$R = \sigma a(rT_s^4 - T^4) , \qquad (V.9)$$

where at these temperatures the emissivity of the surface has been taken equal to the absorptivity. Assuming the energy absorbed by the surface is fed to a Carnot engine operating between the temperature T and the temperature $T_e$, the temperature of the earth's atmosphere, 293$^\circ$ K, the net rate at which work may be done per unit area of the absorbing surface $\dot{W}$ is then given by

$$\dot{W} = \sigma a(1 - T_e/T)(rT_s^4 - T^4) . \qquad (V.10)$$

Differentiating with respect to T and setting the result equal to zero, the temperature $T_o$ yielding the maximum rate of work is given by

$$4T_o^5 - rT_eT_s^4 - 3T_eT_o^4 = 0 . \qquad (V.11)$$

The rate of work has been maximized and not the efficiency, a point stressed in Chapter II. To an excellent approximation, Equation V.11 yields

$$T_o \approx T_s(rT_e/4T_s)^{1/5} . \qquad (V.12)$$

The overall efficiency of the mirrors, surface and Carnot engine for the maximum rate of work $\eta'$ may be obtained from Equation V.10 and V.12 for T by dividing by Equation V.8; thus,

$$\eta' = ar(1 - 4q)(1 - q) , \qquad (V.13)$$

where

$$q = [(T_e/4T_s)^4/r]^{1/5} = 0.030 \ r^{-1/5} . \qquad (V.14)$$

For a few values of r and a the overall efficiency in percent is:

| r   | 1  | 0.9 | 0.75 | 0.5 |
|-----|----|-----|------|-----|
| a   |    |     |      |     |
| 1   | 85 | 76  | 63   | 42  |
| 0.9 | 76 | 69  | 57   | 37  |
| 0.8 | 68 | 61  | 50   | 33  |

The temperature of the surface for the ideal case of $r = 1$ and $a = 1$ is $T_o = 2400°$ K, the other cases yielding smaller temperatures.

The need to store solar energy could be removed if the energy developed could be transmitted from the sunlit side of the earth to the dark side. This might be done, in principle, by connecting solar power plants around the world with a superconducting transmission line; such a solution does not appear feasible. Solar power plants might also be connected by converting electrical energy to microwaves which could be beamed to mirrors (or receivers and senders) in orbit about the earth which in turn could reflect the energy around the world to receiving stations. Conversion of electrical energy to microwave energy can be achieved with about 85 percent efficiency, and from microwave to electrical with about 70 percent efficiency. The energy might, thus, be beamed around the earth at an overall efficiency of perhaps 40 percent, which exceeds the efficiency of some transmission lines over only a few hundred miles.

Another way to avoid the problem of storing energy and also to keep the solar power plant functioning twenty-four hours a day would be to launch mirrors as satellites which could stay fixed above a point on the equator at an altitude of 34,900 km. A sufficient number of such mirrors with the proper rotation and orientation could focus light onto a large area of a few hundreds of square miles where the sunlight could be converted to electricity day and night. The wide scatter of incoherent light would yield some additional illumination over a wider area than needed. The value or lack of value of this *unwanted* radiation would have to be considered. The amount of power received need not be limited to the normal solar flux. The initial costs of the power plant should be comparable to a coal burning plant of the same power except for the cost for the space mirrors.

In outer space about twice as much solar energy flux is available as on the surface of the earth, the insolation being about half the incident solar flux above the atmosphere. In addition the energy is available twenty-four hours a day, so that about four times the energy is available in outer space as compared with the surface of the earth per unit of area normal to the solar rays. In addition no storage of energy is needed. It has been suggested that the energy from a solar energy converter placed in orbit be beamed to the earth as microwaves. If solar cells are used at an efficiency of 10 percent and the conversion and transmission is 60 percent efficient, the overall efficiency is 6 percent. But including the factor of 4 this is equivalent to 24 percent compared with earth-based converters. This scheme may be economically feasible today.

## Energy Storage

A major problem in the utilization of intermittent energy sources, such as sunlight and wind, is energy storage.

Energy stored mechanically can be recovered with 100 percent efficiency. The drawbacks of mechanical storage are high initial cost, large bulk and lack of motility. Energy can be stored in raised weights (such as water pumped to a greater height), deformation of a solid (such as a stretched steel spring), compressed gases and kinetic energy (such as a rotating flywheel). Power utilities are now constructing storage facilities using raised water. Flywheels are sometimes used to store energy in order to produce large, but intermittent power output.

Energy storage in static electromagnetic fields is very small per unit volume and is extremely costly. If a static electric field of $10^7$ volts/m could be attained before breakdown in a dielectric with a dielectric constant of 10, an energy density of $\epsilon E^2/2 = 4000$ joule/m$^3$ = $10^{-6}$ kilocal/cm$^3$ would be obtained, which, while large for electric fields, may be compared with the oxidation energy of a cubic centimeter of sugar (1.6 gm of sucrose) of 6 kilocalories. To store energy in a static magnetic field it may be assumed that superconductors are used to maintain the current without losses. The maximum magnetic field that the superconducting state can tolerate is about $10^4$ gauss, yielding an upper energy density of about $\mu H^2/2 = 5 \times 10^6$ joule/m$^3$ = $10^{-3}$ kilocal/cm$^3$. This represents about a thousand times more energy per unit volume than can be stored in an electric field, but it is still small compared with chemical storage. In outer space where volume limitations are not as great energy stored in static electric and magnetic fields might be more practical. Energy stored in alternating electromagnetic fields (inductive or radiative) decays very rapidly due to losses on the boundaries.

Storage of thermal energy requires an investment in mechanisms to circulate a coolant and in large quantities of material. The heat capacity of the material should be high and a phase transition involving large amounts of energy would be desirable. Because thermal energy is usually stored at low temperatures, the efficiency of conversion back to work is poor. In addition thermal energy lacks mobility. Some extraneous heat losses may be expected.

Energy stored as chemical energy has a number of attractive features. It provides the greatest amount of energy per unit volume now known. For example, it would be necessary to raise one cubic centimeter of water 2,520 kilometers to equal the oxidation energy in one cubic centimeter of sugar of 6 kilocalories. The energy stored per unit mass is also large, thereby providing the best source of stored energy for automobiles, trains, trucks, airplanes and rockets. The efficiency of conversion from electrical or mechanical energy to chemical energy and then back to electrical energy is poor, or the process is costly. However, in principle, overall storage efficiencies of about 60 percent should eventually be economical. A storage battery (electrical-chemical-electrical) gives an overall efficiency of about 50 percent. A fuel cell run on hydrogen (electrical-hydrogen + oxygen-electrical) gives an overall efficiency of about 60 percent. For large power needs storage batteries and fuel cells are too expensive. The electrolysis of water to hydrogen and oxygen which is stored and then burned in a conventional steam power plant could yield an overall efficiency of about 30 percent. However, the storage of gaseous hydrogen and oxygen is awkward and expensive. Chemical processes, being statistical in nature and therefore irreversible at finite rates, can never approach the 100 percent efficiency of mechanical storage; yet the high concentrations of energy stored makes chemical storage the most attractive for future development.

If it were possible to use atoms with metastable states involving greater transition energies than normally associated with chemical processes, great improvement would be achieved. For example, transitions between the electronic states in an atom or the nuclear states in a nucleus could give energy concentrations hundreds or millions of times greater. Such concentrations would be of particular value in rocketry where the energy per unit mass is important. The preparation of such energized metastable states and mechanisms for releasing their energy have yet to be developed.

### Sources of Mechanical Energy

All practical sources of mechanical energy found on the earth, with the possible exception of tidal energy, derive their energy originally from sunlight. Results are summarized in Table II.

WATER POWER. The gravitational energy released by water flowing to lower elevations has been used as a source of energy since prehistoric times. Prehistoric man undoubtedly used rivers to transport himself and goods downstream. Tibetan prayer wheels have probably been turned by water power since the use of writing and the invention of paper, about 1000 B.C. Vitruvius in 27 B.C described a Roman water mill for grinding wheat, a device that had probably already been in existence for some time. By 1000 A.D. water mills for grinding grain were common enough to indicate that most grain was ground in such mills. The construction of large dams and the use of electric generators and power lines have permitted the harnessing of a large amount of water power in recent years.

If it is estimated that 40 percent of solar insolation goes into evaporating water and raising it into clouds, there are about $4 \times 10^{16}$ watts expended in the hydrological cycle. Only about 1 percent of this energy is involved in raising the water into clouds, the rest being energy necessary to evaporate the water. At cloud level there are then about $4 \times 10^{14}$ watts of power available. By considering the amount of rainfall and the elevations over the earth the water power available at ground level has been estimated by Ayers and Scarlott (1952) to be $3 \times 10^{13}$ watts, no more than 5 percent of this being practical, or $1.5 \times 10^{12}$ watts. For a world population of 4 billion the potential practical water power per person is only 400 watts, which compares with an actual production in 1970 of about $8 \times 10^{10}$ watts, or about 20 watts per person.

Water power, being a mechanical source of energy with mechanical storage, is one of the most desirable sources of energy possible. The amount is limited, however, and all available power should become harnessed with no further increases possible.

The cost of water power is only the initial outlay and maintenance costs. Development of water power then depends primarily upon the economic expectation half-life.

WIND. Wind has been used since prehistoric times for the transportation of men and goods across bodies of water in sailboats. This utilization of a mechanical source of energy to transport goods was a major stride forward in the early evolution of man's economic system. It provided man with the capacity to maintain order at a level far above anything known before. Windmills were probably used as early as 1000 A.D. By the thirteenth century windmills had become fairly common (Reynolds, 1970). Windmills were used to grind grain and raise water. Holland's famous windmills made it possible to recover large areas of fertile land from the sea. Today windmills have been largely replaced by gasoline engines or electric motors.

To obtain a rough estimate of the total energy in the earth's surface winds it may be estimated that the atmosphere up to 150 meters moves with an average speed of about $v = 400$ cm/sec. From the surface area of the earth, $A = 5.08 \times 10^{18}$ cm², and the density of air, $\rho = 1.29 \times 10^{-3}$ gm/cm³, the kinetic energy of the earth's atmosphere up to $h = 150$ meters is approximately

$$hA \, \rho \, v^2/2 = 8 \times 10^{24} \text{ ergs} . \qquad (V.15)$$

If it is assumed that this energy is dissipated each day (its purpose being to convect aloft the solar energy trapped next to the earth), the power that might be available is $9 \times 10^{12}$ watts

(Ayers and Scarlott estimates $3 \times 10^{13}$ watts [Ayers and Scarlott, 1952]). Perhaps 5 percent of this energy could be actually harnessed, or $5 \times 10^{11}$ watts. For 4 billion people this is 100 watts per person, about one-fourth as great as water power.

The efficiency with which energy may be extracted from wind depends upon the wind velocity. The power developed from wind varies as the cube of the wind velocity $v^3$, since the power depends upon the kinetic energy passing per unit time through the wind turbine. For winds much less that about 1000 cm/sec little energy can be extracted, and for winds greater than about 1500 cm/sec wind turbines are unable to handle the large load (Putnam, 1948). If energy is to be delivered at a steady rate, expensive equipment must be provided to store energy over long periods due to the erratic nature of the wind. A number of wind power plants might be integrated with an electrical network over a large area to help smooth out erratic fluctuations; however, weather patterns persist over many hundreds of miles, so large fluctuations would still remain.

The practical economic utilization of wind power is still largely in an experimental stage. It appears that wind power will eventually be used at particularly favorable sites where the wind is strong and reasonably constant. Because of the intermittent character of the wind, windmills should be best suited for jobs that depend upon the total work done rather than upon a continuous supply of power such as pumping water. But due to high initial cost and low efficiency small windmills have been largely replaced by the more economic gasoline engine or electric motor even for pumping water.

TIDES. The periodic rise and fall of the ocean level is a source of useful power. Since prehistoric times the outgoing tide has been used to transport men out to sea. The first water-pumping system for the city of London was run by a tide wheel set up in the old London Bridge prior to 1824. A full-scale power plant was put into operation in 1966 at Rance, France, with a power capacity of $8 \times 10^7$ watts. Power is generated only during a four- to six-hour period of each twelve-hour tidal cycle. A number of sites have been proposed for tidal power plants where the twice daily difference in height between high and low tides is around

7 to 10 meters averaged over the year, such as the Severn River in southern England, the bay of Fundy in Canada, the Passamaquoddy Bay between Maine and New Brunswick, Mont St. Michel in France, and the San Jose and Deseado Rivers in Argentina.

Due to the rotation of the earth in the presence of the moon and the sun there is a periodic gravitational force acting on the water in the earth's oceans which produces the rise and fall in the ocean level, or tides. The potential energy of the water at high tides compared with the potential energy at low tide indicates the potential energy in tides. Some of the energy is dissipated as friction with the ocean shores and turbulence in shallow waters; this energy must be supplied from the total mechanical energy of the system of the moon, earth and sun. There is evidence that the rate of rotation of the earth is slowly decreasing, or the earth day is slowly increasing. The approximate energy in twice daily tides, assuming an average height of $h = 30$ cm from high to low tide, is

$$\rho \, \text{Agh}^2 = 3.6 \times 10^{24} \text{ ergs} , \qquad (V.16)$$

where $\rho = 1$ for water, $A = 4.06 \times 10^{18}$ cm$^2$ is the area of the oceans, and $g = 980$ cm/sec$^2$ is the acceleration of gravity. If it is assumed that roughly half of this energy is in principle available, the total conceivable tidal power becomes $2.1 \times 10^{12}$ watts.

Practical power may be obtained from tides by allowing water to flow into prepared basins at high tide and to flow out at low tide. A two-basin arrangement provides a steadier source of power than a single basin. The total energy that can be extracted is decreased by requiring greater time continuity in the power delivered. The ratio of practical power to the total conceivable power may be roughly taken as equal to the area of shallow tidal basins to the area of the earth's oceans, or a factor of about $10^{-3}$. The amount of practical tidal power that is potentially available is roughly estimated to be $2 \times 10^{11}$ watts. (See the conference volume edited by Gray and Gashus, 1972.)

WAVES. The energy in water waves (Kinsman, 1965) comes from the wind and, therefore, comes indirectly from sunlight.

Waves are generally too small to yield practical power; the short height from crest to trough represents too small a head of water to operate a useful power device. Waves in surf may be generally higher, but they are still too small. The high waves that sometimes accompany storms are too infrequent to be of value.

If it is assumed that water waves possess 1 percent of the energy of the wind (up to 150 meters) the total power available, in principle, in waves would be about $9 \times 10^{10}$ watts.

BAROMETRIC PRESSURE. The barometric pressure of the atmosphere changes from day to day. When the air pressure rises it can compress a bellows which has a fixed pressure inside, thereby doing work; when the air pressure falls the bellows can again expand. This is an extremely small effect; differences in pressure equivalent to a column of water only as high as 25 cm can ordinarily occur. In addition the pressure changes very slowly, so that the amount of power available is truly negligible. Nevertheless, clocks have been constructed that utilize this source of mechanical energy.

OCEAN CURRENTS. In the oceans there are currents, such as the Gulf Stream, that carry large amounts of water. This flow contains mechanical kinetic energy which, in principle, might be extracted for useful purposes. In particular, the energy could be extracted by using an immersed water wheel or turbine. Ocean currents have been used since prehistoric times to aid in the transportation of men and goods on rafts or boats, but there has never been an attempt to extract the power from ocean currents.

Since the precise extent of ocean currents is still not known, it is difficult to estimate the total kinetic energy they contain. Making an order of magnitude estimate, the kinetic energy of the hydrosphere may be taken as comparable to the kinetic energy of the atmosphere (up to 150 meters); thus, $9 \times 10^{12}$ watts of power may, in principle, be potentially available from ocean currents. In the Gulf Stream alone the amount of kinetic energy passing a particular point per unit time is considerable. Assuming the Gulf Stream off of the East Coast of the United States transports $V = 7.5 \times 10^{13}$ cm³/sec of water at a velocity of $v = 120$

cm/sec (Sverdrup *et al.*, 1942; Stommel, 1965), the rate that kinetic energy flows by a particular point is then

$$\rho \, V \, v^2/2 = 5.4 \times 10^{10} \text{ watts} . \qquad (V.17)$$

This large energy is due to the large mass of water that is flowing, the energy per unit time mass being quite small, 7200 ergs/gm. If one half of the energy passing through a turbine placed in the Gulf Stream can be converted to useful power and if 10 megawatts of power are desired, then the diameter of the turbine required is given by

$$d = 2(2 \times 10^{14}/\pi \, \rho \, v^3)^{\frac{1}{2}} = 1.21 \times 10^4 \text{ cm.} \qquad (V.18)$$

Such a large turbine, while not impossible, appears to be uneconomical.

Even though ocean currents represent firm power, the underwater environment for the power plant is awkward.

SEISMIC ENERGY. The energy in earth movements and seismic waves may seem large when viewing the damage that can be produced by earthquakes. This terrible damage, however, is produced in a very short time by very rapid or impulsive motion. When averaged over long periods of time the power released by earthquakes is not large. It has been estimated to be about $3 \times 10^{10}$ watts (Gutenberg, 1957) which is less than the water power actually consumed in 1970. There is no known way to harness seismic energy. The energy in seismic waves probably comes from the convection produced by the fossil heat left in the earth from its original formation from a dust cloud; convection in the mantel produces ruptures and friction in the crust.

## Geothermal Energy (See Tables II and III for a summary.)

VOLCANIC HEAT AND HOT SPRINGS. When the earth's surface buckles and folds or shifts longitudinally along a fault, seismic energy may be converted into thermal energy. The thermal energy thus developed may produce hot spots, molten rock and volcanoes along such earth faults or folds. Because of the very low thermal conductivity of most rock, volcanic heat may accumulate over the ages, and a considerable amount of geothermal energy may become stored in a localized region of the earth's crust.

Despite the high temperatures, and therefore the potentially high efficiency of conversion to work, the energy in molten rock and hot earth in and around an active volcano does not usually constitute a practical source of power. If the heat were taken from a small volume of a volcano, the heat from the remainder of the volcano would not flow into this small volume fast enough to provide any significant energy flux because of the very low thermal conductivity of rock, molten or solid. Consequently a very large and uneconomical volume of the volcano would have to be honeycombed with thermal conductors or water pipes to provide a reasonable flux of energy.

Practical sources of volcanic heat usually depend upon naturally occurring hot springs, geysers or blow holes that eject steam under pressure. The hot water issuing from such wells comes from underground channels that establish a large area of contact with the hot volcanic rocks below the surface. This volcanic heat need not arise from a currently active volcano, since the heat can remain for ages stored in the ground having been created by a formerly active volcano.

The flow of fossil heat from the interior of the earth ($3 \times 10^{13}$ watts) creates convection in the mantel which in turn is responsible for seismic activity and volcanism, so that it can be estimated that the rate volcanic heat is generated is some small fraction of the rate that heat is conducted from the interior. Assuming 1 percent, it is estimated that volcanic heat is generated at a rate of about $3 \times 10^{11}$ watts. If it is assumed that hot springs or geothermal heat of volcanic origin is one third of this and is available at about 66° C, or 340° K, then for an atmospheric temperature of 293° K less than 18 percent of this energy, or only about $10^{10}$ watts, is potentially available as electrical energy.

The economic feasibility of harnessing the energy from a particular hot spring or steam well depends upon the temperature of the water, the volume of the flow, and in some cases upon the pressure of the steam.

At the present time geothermal energy is being converted to electrical power at four sites: Larderello, Italy (since 1904) where $37.2 \times 10^7$ watts of electrical power are being generated; Wairakei, New Zealand (since 1958) where $19.2 \times 10^7$ watts of electrical power are being generated; Geysers, California (since 1960) with $8.2 \times 10^7$ watts; and Japan (since 1968) with $3.1 \times 10^7$ watts. Iceland has plans for the utilization of its abundant supply of geothermal energy for the development of electrical power, in addition to its current utilization of such energy for space heating. There are a few other sites that might eventually be utilized for geothermal energy, but the total power potentially available from all such sites is quite small when compared with other sources of power. (See Hammond, 1972, for a brief review.)

EARTH HEAT. As one progresses down into the earth the temperature rises by approximately 1° C for every 30 meters of depth. This gradient indicates an energy flow of about $3 \times 10^{13}$ watts from the interior of the earth to the surface. The source of this heat is fossil heat trapped during the formation of the earth from a primordial cloud of gas and dust. A small part of this heat may also arise from the decay of radioactive elements.

If the mean internal temperature of the earth is taken to be at least 500° C and the heat capacity is assumed to be 0.2 cal/gm (equal to that for basalt), then the earth contains on the average at least 100 calories per gram of trapped thermal energy. The total thermal energy amounts to at least $2.5 \times 10^{37}$ ergs, which represents a vast store of energy which might be partially tapped.

A power plant to utilize the earth's heat at the rate at which it normally flows to the surface, $6 \times 10^{-6}$ watt/cm$^2$, needs to extract heat over a circular area of 14 km radius in order to collect 40 megawatts of thermal power. If the power plant is to deliver 10 megawatts of electrical power, the heat should be available at about 240° C. This temperature is to be found at a depth of about 8 km. By spacing an appropriate number of nuclear bombs of the right size at 8 km depth and detonating them simultaneously it appears that it should be economically possible to create a sufficiently large region porous to water circulation, so that the desired amount of heat at the desired temperature could be obtained by circulating water under pressure. In particular, this technique would create artificially a situation similar to the one found naturally where hot springs are used for electric power.

Power plants using the earth's heat would be pollution free and could be located anywhere without disturbing the surface ecology.

THERMAL ENERGY IN THE OCEAN. Below about 2 km the ocean is at 4° C. At this temperature water is the most dense. The surface of the ocean may be hotter or colder than this temperature, so that in principle a heat engine could be operated between the two temperature reservoirs at the top and below. More than half of the ocean has an upper temperature greater than 20° C (but never greater than 32° C). At about 1 km the temperature drops to 10° C. If the first 0.5 km of ocean is taken as a hot reservoir at 18° and the cold reservoir is taken as the water below 2 km at 4°, the available energy in the upper 0.5 km is then 14 calories per gram times the number of grams. Assuming an area of $3.0 \times 10^{18}$ cm² of warm ocean, this amounts to about $8.8 \times 10^{31}$ ergs of stored energy. The Carnot efficiency between these two temperatures is 5 percent. If a real heat engine were to give only one-fifth of this value, it would still give a mechanical work equivalent of $8.8 \times 10^{29}$ ergs of stored energy. This is a little more energy than contained in all of the coal stored in the earth which has a mechanical work equivalent (at 30 percent conversion) of about $6.0 \times 10^{29}$ ergs. In addition, the energy from the ocean is much more readily available than coal, since it lies on the surface of the ocean.

The problem of tapping this huge source of energy is the problem of development of a technology for handling large quantities of water at small temperature differences of about 14° C. Georges Claude, the French inventor, was the first to do experiments to try to develop this energy source. A full-scale pilot plant study off of the coast of West Africa was made by the French corporation Energie des Mers in 1953 (Eng. News Rec., 1953). Hot water at 31° C was to be taken from a lagoon and cold water at 6° C was to be taken from an ocean deep 4 miles off the coast. The hot water was to be vaporized and sent through a vacuum turbine to generate power. The full-scale plant was to produce $7 \times 10^6$ watts of power at an initial outlay of 2.5 million dollars. Evidently the venture was not economically attractive at that site at that time, and it was discontinued. The ocean provides a pollution-free energy source which even subtracts from the thermal pollution. Eventually technology will probably develop to the point where this huge source of energy will be tapped.

It is of interest to inquire where this energy comes from. After a sufficient lapse of time the ocean should come into thermal equilibrium if there were no process to continually maintain the temperature difference between the top and below the surface. The difference is maintained by ocean currents on the ocean floor carrying cold water from the polar regions toward the equator while surface currents carry warm water from the equatorial regions toward the poles. The rate that energy is supplied to maintain the temperature difference is not large, since thermal conductivity is small and there is no mixing from the top below, the water below being denser. The only energy flow that can upset the balance is the flow of heat from the earth's interior. Thus, solar energy, driving the ocean currents, refrigerates the bottom of the ocean by removing about 75 percent of the earth's heat, or $2 \times 10^{13}$ watts. It may then be roughly estimated that at least this amount of energy is continuously available without subtracting any energy from the amount stored.

OTHER SOURCES WITH SMALL TEMPERATURE DIFFERENCES. The temperature of about 8° C between day and night for air near the surface of the earth might conceivably be a source of energy, since it may represent perhaps 2 percent of the solar energy flux to the ground. Unfortunately, there is no way to store such low utility energy or to handle the large volume of air that would be needed. No real irreversible heat engine could operate successfully with such a small temperature difference.

On mountain tops the temperature is colder than in the valleys below. It is, thus, conceivable that a heat engine could operate between these two temperatures, the reservoirs being perhaps two lakes at the two altitudes. The temperature differences possible are small, however, and the practicality of such a scheme is the same as using the thermal energy from the ocean.

Since the earth maintains a relatively constant temperature a hundred or so feet underground, it is conceivable that the seasonal variation in atmospheric temperature might be used as a source of power. The constant temperature reservoir could be water from a well, and the seasonal source could be a lake. Here again the temperature differences are, in general, too small to permit a reasonably profitable conversion of heat to electricity. Without thermal storage spring and fall would be awkward periods of zero or little power.

Waste heat from conventional and nuclear power plants, furnishes a source of considerable energy but at a small temperature difference.

Thermal sources with small temperature differences cannot be discounted as uneconomical where space heating is concerned. Heat pumps in conjunction with such sources can yield economical heating and cooling. The main problem is the availability of such sources to the consumer.

## Physical-Chemical Energy

Physical-chemical energy sources arise from such processes as freezing, evaporation, dissolving a solute in a solvent, and other such changes in state or phase that do not ostensibly involve changes in chemical structure.

ENERGY FROM THE EARTH's DRY ATMOSPHERE. The air immediately above the surface of the earth is usually not saturated with water vapor. By evaporating water into this dry air it is possible to reduce the temperature of a reservoir to the wet bulb temperature. A heat engine may then be operated, in principle, between a hot reservoir at the dry bulb temperature and a cold reservoir at the wet bulb temperature. This constitutes a significant source of energy. For example, the dry atmosphere is used to produce about 2 percent of all of the electric power generated in the world today, about $9 \times 10^9$ watts. Cooling towers that evaporate water are used in conjunction with power plants that use coal, gas, oil or nuclear fuel. Cooling towers that evaporate water are able to maintain a colder temperature than dry cooling towers. The power made available by the dry atmosphere is then given by the difference between the power delivered using cooling towers that evaporate water and the power delivered when using cooling towers that do not evaporate water.

Plants and animals also employ the earth's dry atmosphere

as an energy source. Land plants take up a large quantity of water from the ground which is then evaporated from the leaves. In this way work is done to bring essential compounds up from the soil to be deposited in the plant. Since the quantity of water evaporated generally exceeds the amounts needed to supply mineral nutrients, plants apparently use the earth's dry atmosphere as an energy source in addition to direct sunlight. Some animals use the evaporation of water to maintain a constant body temperature during hot weather. The evaporation of water then makes available additional power in precisely the same way that cooling towers provide additional power to steam-electric power plants.

Water is being continually evaporated from the ocean, lakes, forests, etc., by the direct action of sunlight and warm dry air. The atmosphere is maintained dry next to the surface by moist air being convected aloft. After the moisture condenses out as rain, the dry upper air is then recycled down to the surface of the earth. This convection is part of the process whereby thermal energy trapped by the greenhouse effect is carried aloft to be radiated into space. Energy absorbed by water (540 cal/gm) when evaporated at the earth's surface is relinquished in clouds during condensation. If $4 \times 10^{16}$ watts are involved in this hydrologic cycle, the dry, ground-level air might provide a source of energy equal to perhaps 1 percent of this, or $4 \times 10^{14}$ watts. Assuming that wet bulb and dry bulb temperatures ordinarily differ by about 15° C, a Carnot engine could supply about 5 percent of this energy as work, or $2 \times 10^{13}$ watts. The temperature difference of 15° C is too small, however, for the development of practical power from dry air and water alone (i.e. not in conjunction with a steam power plant). It is, however, economical for space heating where it is frequently used.

ENERGY FROM ICE. A heat engine placed upon an ice floe in the Arctic Ocean could use the unfrozen water below the floe as a source of water to freeze. The hot source would then be at − 2.5° C, the temperature of freezing sea water. If the cold source were at − 30° C, the temperature of the atmosphere, then an ideal heat engine would operate with an efficiency of 10 percent. Thus, 8 calories out of each 80 calories to freeze a gram of water would be available for useful work on the ice floe.

It is also possible to use ice as a cold source. The ice from Greenland or the Antarctic could be floated, in principle, to warmer parts of the earth where a heat engine could operate using warm water or the atmosphere as a hot source. Again it may be estimated that about 10 percent of the 80 cal/gm of ice could be ideally converted to useful work.

The freezing or melting of ice does not constitute a major source of power, even if it were practical. It may be economical, however, for space heating.

## Chemical Energy from Current Life Processes

Solar energy is converted to chemical energy by the process of photosynthesis in green plants. The net result is to reduce carbon dioxide in the presence of water to produce carbohydrates and oxygen; thus,

$$n\ CO_2 + m\ H_2O \rightarrow C_n(H_2O)_m + n\ O_2 ,$$

where n and m are integers. The energy available upon oxidation is then approximately the oxidation energy of carbon, $4.0 \times 10^{12}$ ergs/mole.

To estimate the maximum amount of energy that might be reasonably obtained by the use of photosynthesis, it may be estimated that about 2 percent of the incident sunlight (insolation) upon a well-watered and cultivated field of plants, such as wheat, can be converted to chemical energy stored in the plant. It might seem that 2 percent is small compared with what might be ideally expected (perhaps as high as 50 percent) by direct physical means, but approximately 10 percent of the incident light is reflected from the leaves; about 75 percent of the incident light is of wavelengths not suitable for photosynthesis; the growing season (averaged over the earth) is only about 50 percent of the year; about 30 percent of the gross energy stored is lost in respiration to maintain the plant; and the chloroplast itself has a conversion efficiency of about 75 percent. Considering all of these factors, the ideal maximum efficiency is only about 6 percent, which makes the practical maximum of 2 percent appear reasonable. If one-tenth of the surface area of the earth can be used to grow plants, this yields a rate of supply of chemical energy equal to $1.8 \times 10^{14}$ watts. This may be compared with the observed rate which is estimated to be $4 \times 10^{13}$ watts. If the chemical energy is converted to electrical energy in fuel cells with an efficiency of 70 percent, the practical maximum electric power that can be supplied by photosynthesis is $1.3 \times 10^{14}$ watts. This amounts to 30,000 watts per person for a population of 4 billion. A typical daily consumption of 2000 kilocalories of food per person is equal to 100 watts.

Wood has always constituted a major source of fuel. Wood has been used since prehistory for space heating and for the manufacture of essential compounds. In the early days of the steam engine, wood was the primary source of power. Even as late as 1950 about $8.5 \times 10^{10}$ watts of power were derived from wood, representing about 7 percent of the world's energy consumption.

Plant energy is still the sole source of energy for animals. Humans as well as other animals still depend upon plants to supply the compounds and fuel for their bodies. Viewed as heat engines, animals and humans deliver useful work using the chemical energy provided by plants. Considering the completely biological sun-plant-animal system for converting solar energy to useful work, the system is far less than 1 percent efficient. Animals can at best convert about 20 percent of the food energy assimilated to useful work, neglecting the energy that has to be consumed to attain maturity and the energy expended during necessary periods of inactivity. If it is roughly estimated that 2 percent of all the chemical energy stored in plants is actually converted to mechanical work by animals, then about $8 \times 10^{11}$ watts of mechanical power is delivered to the biosphere. See Table II for a summary.

One of the advantages of the photosynthetic conversion of solar energy is that the chemical energy obtained may be stored to be available over periods and seasons when the sun does not shine. The energy density of this stored energy is large, requiring little volume or mass for storage. Its high energy density makes it suitable for motile devices, animals and machines. In addition, there is no capital expense to manufacture the device (the green plant) that converts and stores the energy. The raising of plants and animals for the conversion of sunlight to useful work has been carried on economically since prehistory. Only in very recent times, since the industrial revolution, has the utilization of fossil energy by machines largely replaced the sun-plant-animal complex for the delivery of mechanical work. Due to very basic inefficiencies, however, such as the conversion of sunlight to chemical energy at an actual efficiency of less than 2 percent, ordinary biological systems will eventually be replaced (as explained in Chapter IX).

## Chemical Energy in Fossil Fuels

While most of the carbon in an ecosystem is recycled, there can be a small rate of deposition of detrius in a reducing environment that can accumulate over the ages to form large fossil deposits. These deposits contain large stores of chemical energy.

OIL. Petroleum has been used since prehistoric times for space heating, cooking and lighting. The Chinese were drilling oil wells as early as 1000 B.C. It has only been recently, however, that the energy from oil has been extensively converted to useful mechanical work. In 1968 the world production of crude petroleum was equivalent to $7.2 \times 10^{26}$ ergs of energy (Statistical Office of the United Nations, 1970). Oil yields about one third of the total raw energy (apart from photosynthesis) produced in the world today ($1.74 \times 10^{27}$ ergs/yr) for a rate of supply equal to $229 \times 10^{10}$ watts.

Ideally about 65 percent of the energy in oil can be converted to electrical energy, a figure approached by fuel cells. Thus, ideally about $149 \times 10^9$ watts of electric power were potentially available in 1968 from crude oil. For a world population of 4 billion this rate would provide 370 watts per person. In actual practice refining and transportation of oil use about 20 percent, and the heat engines actually used, primarily gasoline engines, are less than 30 percent efficient. The actual rate of delivery of mechanical power (apart from electric power) from oil may be estimated to be about $35 \times 10^{10}$ watts in 1968. Electric power amounted to about $6.8 \times 10^{10}$ watts.

The total reserves of oil in the ground are limited, so that eventually the available oil will be consumed. Any estimate of the reserves of oil in the ground that can be eventually recovered is largely speculative. According to usual estimates the total reserves of oil amounts to about $1.2 \times 10^{29}$ ergs of energy. At the 1968 rate of using oil the earth's oil reserves would last about two hundred years. Since the world population and economy are expanding rapidly, this oil reserve will probably be depleted long before two hundred years have elapsed.

COAL. Coal was first mined and used in China in about 1100 B.C. The early use of coal was in ceramics, metalurgy, space heating and cooking. It has only been since the invention of the steam engine that the energy in coal has been converted to useful mechanical work.

In 1968 the world production of coal (including small production of lignite and peat) amounted to $6.45 \times 10^{26}$ ergs which is a delivery rate of $204 \times 10^{10}$ watts. This is a little less than the energy in petroleum and amounts to about one third of the total energy production. Ideally at 65 percent efficiency this would yield about $130 \times 10^{10}$ watts of potential electric power. A population of 4 billion would receive 510 watts per capita of raw power in coal at the 1968 rate. About 58 percent of the coal was used at about 30 percent efficiency to yield $35.5 \times 10^{10}$ watts of electric power.

The total reserves of recoverable coal in the ground are quite large, but still finite, so that eventually the supply of coal will become exhausted. According to usual estimates the total world coal reserves amount to about $2 \times 10^{30}$ ergs. At the 1968 rate of using coal the earth's coal reserves will last about one thousand years. The expansion of the world population and economy will, however, exhaust the coal long before the one thousand years have elapsed. The total reserves of coal still in the earth exceeds the combined reserves of all other fossil fuels by about a factor of 5, making coal in the long run, the most important of the fossil fuels.

NATURAL GAS. In 1968 the world produced $3.38 \times 10^{26}$ ergs of energy in natural gas for a rate of $107 \times 10^{10}$ watts which would give 270 watts per person to a population of 4 billion. About 17 percent of this energy went into producing electricity at an efficiency of about 30 percent; or $5.5 \times 10^{10}$ watts of electrical energy were produced.

The total world reserves are estimated to be about $1.1 \times 10^{29}$ ergs, about the same as the total world oil reserves. At the present rate of consumption natural gas would last about three hundred years.

OIL SHALE. Although the total energy content in the earth's reserve oil shales of about $2 \times 10^{29}$ ergs may be estimated as about one-tenth of the energy in the earth's coal reserves, much less than one-tenth can, in fact, be extracted, since a large amount of energy is needed to retort the shale to obtain the oil. Due to the cost of mining and extracting the oil from the shale, oil shales will not be extensively exploited until the world's oil reserves become largely depleted.

PEAT. Peat is currently consumed on a small scale, particularly for space heating. The total reserve energy in peat is about 1 percent of the reserve energy in coal. At the present rate of accumulation of peat about $48 \times 10^{10}$ watts of power are continuously available.

ASPHALT, TAR, PITCH, GILSONITE, ETC. Highly viscous petroleum deposits currently find no use as energy sources because of the difficulty in handling them. These deposits are of interest because they contain long hydrocarbon molecules that are valuable in paint and other coatings. It may be roughly estimated that the total reserve of tars, etc., equals the reserve of liquid petroleum. When ordinary oil reserves become depleted, techniques for handling tars for energy may be developed.

## Energy in Sulfur and Sulfides

Most of the world's native sulfur is obtained from a few large sulfur domes along the coast of the Gulf of Mexico. Native sulfur can be used to generate useful power by first burning it to sulfur trioxide, $SO_3$, yielding 108 large cal/mole and then dissolving the resulting $SO_3$ in water to produce sulfuric acid, $H_2SO_4$, and another 100 large cal/mole (Hodgman, 1961). The resulting sulfuric acid may be combined with a base to yield a sulfate. This process may be employed in a wet cell battery to yield useful electric current. The energy in forming the sulfate, however, comes primarily from the cation rather than the anion. Thus, a mole of native sulfur will yield 207 kilocalories.

In 1968 the world production of native sulfur was $1.41 \times 10^7$ metric tons (Statistical Office of the United Nations, 1970), which represents an energy production of $3.82 \times 10^{24}$ ergs/yr, or $1.2 \times 10^{10}$ watts. Perhaps about one-fourth of this energy could be converted to electrical energy. Sulfur plays a valuable role in the chemical industry and is not used for its energy content.

When iron sulfide, pyrites, $\frac{1}{2}FeS_2$ is oxidized to sulfur trioxide, $SO_3$ and iron oxide, $\frac{1}{4}(Fe_2O_3)$, 255 large cal/mole of sulfur are released. In 1968 $1.03 \times 10^7$ metric tons of sulfur were produced from pyrites for an energy equivalent of $3.44 \times 10^{24}$ ergs/yr, or $1.1 \times 10^{10}$ watts (see Table III).

It is very likely that sulfide ores are essentially fossil deposits like coal, because most thermodynamic processes that have occurred in the lithosphere probably started with life; sulfide ores are associated with limestones; and pyrites are frequently associated with ancient microfossils. The amount of sulfide ores in the earth has not been estimated, but the fact that sulfur is a common element indicates that there may be considerable deposits.

### Terrestrial Electromagnetic Energy

Terrestrial sources of electric and magnetic energy cannot yield practical amounts of power, but they are considered here for completeness and because similar nonterrestrial sources may not be insignificant.

LIGHTNING. The energy dissipated in an average lightning bolt is quite large, amounting to $10^{17}$ ergs (Chalmers, 1957). If it is assumed that there are about two lightning flashes per year per square kilometer (Golde, 1945), then the energy dissipated in lightning flashes over the earth is about $10^{26}$ ergs/yr, or $3 \times 10^{11}$ watts. This total power is not very large and there is no known way to harness it.

THE EARTH'S ELECTRIC FIELD. Near the earth's surface there is an electric field of about 365 volt/m, the earth's surface being negative with respect to the ionosphere. This electric field produces a total current from the ionosphere to the earth of about 1300 amperes. Assuming the resistance of the earth's atmosphere to be about 200 ohms, the power dissipated by this current is only about $3.4 \times 10^8$ watts. There is no known way to harness this energy. The energy for this atmospheric current is evidently supplied by thunderstorms powered by sunlight.

EARTH CURRENTS. In the surface of the earth there are electric currents produced by thunderstorms, thermal gradients, temporal changes in the earth's magnetic field (telluric currents), and some commercial applications of electric power. None of these sources of electric current gives any hope of yielding practical amounts of useful power with the possible exception of currents produced by chemical processes, geovoltaic currents. For example, an electric potential can be established between a layer of sulfide being oxidized near the earth's surface and an inactive layer of sulfide far below the surface. The total worldwide power available is negligible and the costs to develop such sources would probably be prohibitive.

MAGNETIC FIELDS. The energy in the entire earth's static magnetic field amounts to only $2.5 \times 10^{25}$ ergs, about the energy in

TABLE II.  POWER IN RENEWABLE SOURCES IN $10^{10}$ WATTS

| Source | Total | Practical Potential | Potential for Electric Power | Actual Mechanical Power 1968 |
|---|---|---|---|---|
| *Solar* (sunlight at earth's surface) | 9,400,000 | 900,000 | 200,000 | negligible |
| *Mechanical* | | | | |
| water power | 3,000 | 150 | 140 | 12.4 |
| wind | 900 | 70 | 50 | small |
| ocean currents | 900 | 20 | 15 | 0 |
| tides | 210 | 20 | 10 | 0.008 |
| ocean waves | 9 | very small | 0 | negligible |
| barometric pressure | 10 | very small | 0 | negligible |
| seismic | 3 | 0 | 0 | 0 |
| *Photosynthesis* | | | | |
| total practical | — | 18,000 | 1,800 | — |
| total actual | — | 4,000 | 400 | ~80. |
| wood | 1,000 | 300 | 80 | negligible |
| peat | 5 | 5 | 1 | 0 |
| *Thermal* | | | | |
| hot springs | 30 | 10 | 1 | 0.0679 |
| earth heat | 3,000 | 600 | 150 | 0 |
| seasonal | ? | small | small | 0 |
| day-night | 200,000 | 0 | 0 | 0 |
| *Dry air* | 40,000 | 4,000 | 40 | negligible |

the sunlight falling on the United States in one day. Magnetic storms represent a trivial quantity of energy, the disruption of the ionosphere and radio communications requiring very little energy. The fossil magnetic field preserved as residual magnetization in rocks contains no significant amount of energy. Similarly the magnetic field energy in magnetic ores is negligible.

### Nuclear Energy

Nuclear energy arises from the conversion of mass, m, to energy, E, according to Einstein's mass energy equivalence principle

$$E = mc^2 , \qquad (V.19)$$

where $c = 3.00 \times 10^{10}$ cm/sec is the velocity of light.

FISSION. The process of fission occurs when the nucleus of a large atom—in particular, uranium 235, plutonium 239 and uranium 233—splits into two nuclei of approximately equal mass. The total mass of the two fission fragments is less than the mass of the original nucleus, so that energy is released. The total energy released, apart from neutrino energy of about 10 mev (where 1 mev $= 1.60 \times 10^{-6}$ erg), is $190 \pm 6$ mev per fission. This amounts to $7.8 \times 10^{17}$ ergs per gram of material undergoing fission.

Fission spontaneously occurs at only a very slow rate, but it can be induced by bombarding the nucleus with neutrons. Each fission releases a number of neutrons which may each induce further fission in other nuclei, making a self-sustaining or chain reaction possible (Soodak, 1962). For $^{235}U$ the slower the neutrons the more likely that they will produce fission. A $^{235}U$ reactor contains a moderator to slow the fast fission neutrons down to the required thermal velocities. It may also contain control rods that absorb neutrons (e.g. cadmium rods) to reduce the rate of fission.

The first self-sustaining fission process was achieved at the University of Chicago under the direction of Enrico Fermi in 1942. In 1945 a nuclear bomb based on $^{235}U$ fission was exploded in the New Mexico desert, and another was dropped on Hiroshima. Also in 1945 a nuclear bomb based upon $^{239}Pu$ fission was dropped on Nagasaki. Subsequently fission energy has been used to trigger nuclear bombs based upon fusion energy (see below). The United States now has a fleet of submarines and an aircraft carrier powered by nuclear fission. Russia has nuclear-powered submarines and ice breakers. Since 1954 there has been a continued effort to develop the practical power potential of fission energy. In 1968 the electric power developed in the world from the fissioning of $^{235}U$ was $5.6 \times 10^9$ watts. Electricity produced from $^{235}U$ is today comparable in costs to the more efficient coal-burning power plants.

BREEDING. Reactors have been built using the artificially manufactured element plutonium 239, $^{239}U$. Plutonium is obtained from the plentiful $^{238}U$ by irradiation with neutrons from the fission of $^{235}U$. If the neutrons from the fissioning of $^{239}Pu$ could be used to manufacture more $^{239}Pu$ from $^{238}U$ than are lost in producing fission, a complete 100 percent *breeding* process would be obtained, more fuel being manufactured than consumed. The amount of fuel is then limited to the supply of the plentiful $^{238}U$. In addition to plutonium, $^{233}U$ may be manufactured from thorium, $^{232}Th$, by irradiation with neutrons. Again there is the possibility of a breeding process if the neutrons from the fission of $^{233}U$ manufacture more $^{233}U$ from $^{232}Th$ than is consumed to produce fission.

While feasible, no 100 percent breeding reactors have as yet been constructed for power. Breeding reactors will probably be the main source of energy within twenty years, since they will be able to produce electric power more cheaply than coal.

FISSION ENERGY STORED IN THE EARTH. Due to the spontaneous decay of uranium and thorium and the limited amount of heat flow from the earth's interior it is possible to conclude that uranium and thorium are primarily limited to the earth's crust. In addition, rocks similar in character to the earth's mantel, such as chondrites and moon rocks, have about $3 \times 10^{-3}$ as much uranium and thorium as the crust. The earth's crust contains about 4 ppm (parts per million by mass) of uranium and 16 ppm of thorium (Mason, 1952). The earth's crust has a mass of about $2.4 \times 10^{25}$ gm (Kaula, 1968b), so that the total mass of uranium is $9.6 \times 10^{19}$ gm and of thorium is $3.8 \times 10^{20}$ gm in the earth's crust. Since $7.1 \times 10^{-3}$ part of naturally occurring uranium is $^{235}$U, this means that there are about $6.8 \times 10^{17}$ gm of $^{235}$U in the earth's crust. In energy terms, assuming $7.7 \times 10^{17}$ ergs per gram of material fissioned, there are $5.2 \times 10^{35}$ ergs stored in $^{235}$U, $7.4 \times 10^{37}$ ergs in $^{238}$U (assuming breeding), and $3.0 \times 10^{38}$ ergs in $^{232}$Th (assuming breeding). Since the earth's crust averages only about 17 km thick, all of this energy should be regarded as being eventually accessible.

It is economically feasible to extract the 20 ppm of uranium and thorium from granite for breeding reactors. The fission energy in a gram of granite is about 370 kilocalories. Granite has a heat of formation of about 210 kilocal. per mole and a molecular weight average of about 68.6, which means that about 3.1 kilocalories/gm are needed to completely dissociate granite into its constituent atoms. Since more than 100 times this energy is available, it is clear that only a small fraction of the available energy needs to be used to extract the uranium and thorium. The amount of energy per gram of granite from the fissioning of $^{235}$U, however, is only 0.55 kilocal/gm, so that the feasibility of extracting $^{235}$U for its energy from granite is marginal. A clever separation scheme might make the recovery of $^{235}$U from granite economical, however.

It should perhaps be stressed that each ton of granite, or earth in general, is equivalent to 54 tons of coal when used in breeding reactors, but equivalent to only 0.08 tons of coal for $^{235}$U fission reactors.

THERMONUCLEAR FUSION. The fusion of light nuclei to make heavier nuclei results in a net loss of mass with a resulting release of energy. The sun and the stars obtain their energy from the fusion of light elements, especially hydrogen. Hydrogen occurs as three isotopes: ordinary hydrogen $^1$H, deuterium $^2$H with twice the mass, and tritium $^3$H with three times the mass. The important fusion reactions involving three isotopes are:

$$
\begin{aligned}
^2\text{H} + {}^2\text{H} &\nearrow {}^3\text{He} + \text{n} + 3.3 \text{ mev}, \\
&\searrow {}^3\text{H} + {}^1\text{H} + 4.0 \text{ mev}, \\
^2\text{H} + {}^3\text{H} &\to {}^4\text{He} + \text{n} + 17.6 \text{ mev}, \\
^2\text{H} + {}^3\text{He} &\to {}^4\text{He} + {}^1\text{H} + 18.3 \text{ mev}, \\
^3\text{H} + {}^3\text{H} &\to {}^4\text{He} + 2\text{n} + 11.3 \text{ mev}, \\
^6\text{Li} + \text{n} &\to {}^3\text{H} + {}^4\text{He} + 4.7 \text{ mev},
\end{aligned}
\tag{V.20}
$$

where $^3$He is helium three, n is a neutron, $^4$He is helium 4, and $^6$Li is lithium 6, and 1 mev $= 1.60 \times 10^{-6}$ erg. The possibilities indicated by the first formula are equally likely.

While these processes have been used to obtain energy for nuclear explosions, starting in the late 1940's, no controlled thermonuclear reaction has as yet been attained. The rare, expensive and radioactive (half-life of 12.5 years) tritium can be fused with deuterium at about 30 million degrees temperature, while the cheap plentiful deuterium can be fused with itself only at about 300 million degrees. For comparison, the center of the sun is at about 15 million degrees. At the elevated temperatures required the hot ionized gases can be contained only in magnetic fields, or magnetic bottles. The technology of attaining large enough magnetic fields over a large enough volume of space over a long enough time will probably eventually advance to the point where practical thermonuclear reactors can be constructed; but, considering the fact that a quarter of a century of intensive research has not met with success, it may take many more decades. See the brief review article by Post (1973) for the present state of the art.

FUSION ENERGY STORED IN LITHIUM 6. Considering the last two processes in Equation V.20, it is apparent that two $^6$Li nuclei can be fused in principle to give 10.4 mev energy per nuclei, so that one gram of $^6$Li is equivalent to about $1.66 \times 10^{18}$ ergs, no matter what particular process is being considered. Any source of excess neutrons can be converted to energy by allowing them to be absorbed by $^6$Li. Since $^6$Li is 0.0740 of the naturally occurring lithium, naturally occurring lithium is present in the earth's crust at 65 ppm, and the mass of the earth's crust is $2.4 \times 10^{25}$ gm; there are $1.2 \times 10^{20}$ gm of $^6$Li in the earth's crust. This is then equivalent to $1.9 \times 10^{38}$ ergs of energy stored in $^6$Li in the earth's crust. As discussed above for the breeding of thorium and uranium, this energy is sufficient to make essentially all of $^6$Li in the earth's crust eventually available for fusion.

FUSION ENERGY STORED IN DEUTERIUM. If deuterium is fused with $^6$Li, $^2$H $+ \, ^6$Li $\rightarrow 2^4$He $+ 22.3$ mev, essentially all of the energy available when converting deuterium to helium 4 can be realized, or 12.0 mev is available per $^2$H nucleus. The direct fusion of deuterium, $3\,^2H \rightarrow\,^4He + n +\,^1H$, yields 7.3 mev of energy per $^2H$ nucleus. Since the amount of lithium available can match the amount of deuterium available, the larger figure may be chosen to indicate the energy potentially available, or $5.78 \times 10^{18}$ ergs/gm. The ocean, with a mass of about $1.2 \times 10^{24}$ gm and with $1.56 \times 10^{-4}$ parts of deuterium to ordinary hydrogen, contains $2.0 \times 10^{19}$ gm of deuterium and, therefore, contains about $1.2 \times 10^{38}$ ergs of fusion energy stored in deuterium.

This energy would be sufficient to warrant the extraction of deuterium from the ocean. To dramatically illustrate the vast amount of energy involved it may be noted that each gallon of sea water contains enough deuterium to provide fusion energy equivalent to 210 gallons of gasoline.

OTHER SOURCES OF CONTROLLED NUCLEAR ENERGY. Strontium 90, a fission fragment, emits two electrons in succession to provide 2.78 mev of energy, becoming zirconium 90. This process has a half-life of 25 years. By incorporating such a beta-active nuclide in a semiconducting rectifier, the beta particles produce a cascade of electrons and direct electric power. A more practical utilization is achieved by letting the thermal energy generated drive a thermocouple which in turn produces the desired electrical power. Such a power pack has been included in the first space ship to fly by Jupiter. While such sources may be sometimes convenient, large amounts of power are not involved.

NUCLEAR BOMB EXPLOSIONS. The difficulty in containing and controlling a fusion reaction is entirely obviated in a nuclear bomb explosion. Large amounts of fusion energy may be readily released at little cost. The blast effect of nuclear explosions can be used to move large quantities of earth such as might be required to dig a sea level canal across the Isthmus of Panama. Nuclear explosions might divert or dissipate hurricanes, rejuvenate old oil fields, change the flow of underground water by breaking through certain rock strata, create underground caverns for use as warehouses, and create a reservoir in which to circulate water to utilize the earth's heat.

Unfortunately nuclear explosions, to be economical, must be large and such large explosions are not practical in most applications where explosions are needed. Also the radioactivity released by a nuclear explosion is extremely dangerous to life and it is long lived. Most practical applications would have to contain the explosion underground.

It has been suggested that a nuclear explosion be contained in a large hole in the ground where the thermal energy released could be extracted for subsequent use. Unfortunately, no cavity can remain intact against the over-pressures produced by a nuclear explosion of a size sufficient to be of economic interest.

A fusion device, or *hydrogen bomb*, provides a cheap supply of a large amount of neutrons. The neutrons from such a device could be absorbed in $^{232}$Th or $^{238}$U to produce the fissile nuclides $^{233}$U or $^{239}$Pu which could then be mined and used in a fission reactor. It does not appear that such a scheme is economically practical.

TABLE III.  STORED ENERGY

| Source | Amount Stored ergs | Power in 1968 in $10^{10}$ watts | |
|---|---|---|---|
| | | Rate Produced | Actual Mechanical Or Electric Power |
| *Fossil fuels* | | | |
| crude oil | $120 \times 10^{27}$ | 229 | 41.9 |
| coal and lignite | 2,000 | 204 | 35.5 |
| natural gas | 110 | 107 | 5.5 |
| peat | 20 | small | 0 |
| oil shale | 200 | 0 | 0 |
| asphalt, tar, etc. | 100 | small | 0 |
| *Sulfur and sulfides* | ? | 2 | 0 |
| *Thermal* | | | |
| ocean | 47,000 | 0 | 0 |
| *Nuclear* | | | |
| fission$^{235}$U | $5 \times 10^{35}$ | 6 | 0.56 |
| breeding $^{232}$Th, $^{238}$ U | 4,000 | 0 | 0 |
| fusion $^6$Li | 2,000 | 0 | 0 |
| fusion $^2$H | 1,000 | 0 | 0 |

## Energy Sources in Outer Space

As man and the machine harness greater and greater amounts of energy they will gradually be able to establish themselves in space. Eventually space colonies should become independent of the earth. When this happens man and the machine will have to survive on sources of energy in space.

SOLAR ENERGY IN SPACE. On the surface of the earth the direct utilization of sunlight is not yet practical, but in outer space it is very practical. To best use sunlight in space a spaceship should be in circular orbit about the sun. The earth intercepts sunlight as a circular area of the radius of the earth. This energy is then distributed over a sphere of the radius of the earth, thereby producing a reduction in the radiant flux per unit area to one-fourth of the value in space. In addition, the atmosphere reflects or absorbs almost half of the incident sunlight before it reaches the earth. In space about eight times as much solar energy is thus received per unit area. In space, sunlight represents firm power which does not cease at night nor decrease during cloudy weather. No costly energy storage is required. In space, mirrors may be used to collect sunlight over large areas without distortions produced by gravitational and inertial forces acting upon them. In space no dust, sand, corrosive compounds, hail or rain can damage such mirrors used for collecting sunlight. If the period of rotation is the same as the period of revolution about the sun, there is no need to continually reorient the mirrors as on earth. In space a heat engine can operate with a colder cold reservoir. The cold reservoir can be a radiator shielded from the sun's rays radiating into deep space which might have an effective temperature of about $3.5°$ K.

A spaceship of a fixed size should be placed as near the sun as possible in order to maximize the amount of solar energy available. It would only have to stay sufficiently far away to keep its mirrors or collecting surfaces from getting too hot. An ideal orbit would probably be inside the orbit of Mercury (Wesley, 1966).

As the number of spaceships increases, they will radiate more of their waste heat to each other. Eventually, if a swarm of spaceships becomes numerous enough, they would radiate off into deep space as a collective sphere, in which case they would have to move out from the sun in order to increase their total effective radiating surface. Neglecting this eventuality by considering only 1 percent of the sun's energy to be intercepted by spaceships, the total power available is $3.90 \times 10^{25}$ watts, which is $2 \times 10^7$ times the solar energy intercepted by the earth. In 1970 the per capita consumption of energy was about 11 kilowatts per person in the United States. If this figure is multiplied by a thousand to represent the energy required to support a man in space, the spaceships using 1 percent of the sun's energy could support a population of $3.5 \times 10^{19}$, which is 10 billion times the population of the earth in 1970.

To appreciate just how vast the sun's radiating power actually is, it may be noted that all of the nuclear energy potentially available on the earth as fission and fusion energy in the earth's crust and the ocean, $7 \times 10^{38}$ ergs, would supply the sun with radiant energy for just two days. The sun is such a huge reactor yielding such large amounts of convenient safe power, that no other source in space in the solar system can compare with it.

GRAVITATIONAL ENERGY. The universe contains vast stores of gravitational energy. Purely gravitational systems with moving rigid bodies and point masses can never yield energy for human consumption, because the conservation of gravitational plus kinetic energy prohibits the extraction of any energy from the system. The only way that gravitational energy can ever be used is for the gravitational system to be coupled to a dissipative system. For example, the moon has gravitational energy with respect to the earth. It may be extracted in principle by the following scheme: The moon is broken up a piece at a time; each piece is reduced to zero orbital velocity around the earth; whereupon, it falls to the earth giving up upon impact its energy which goes into thermal energy plus a small amount of kinetic energy to match the speed of the surface of the earth. Neglecting the very small reduction in the earth's rotational energy, the net energy gained is given by

$$GM_mM_e(1/r_e - 3/2r_m) - M_m v_e^2/2 = 4.38 \times 10^{37} \text{ erg },$$

$$(V.21)$$

where $G = 6.67 \times 10^{-8}$ cm$^3$/gm sec$^2$ is the universal gravitational constant, $M_m = 7.35 \times 10^{25}$ gm is the mass of the moon, $M_e = 5.98 \times 10^{27}$ is the mass of the earth, $r_e = 6.37 \times 10^8$ cm is the earth radius, $r_m = 3.84 \times 10^{10}$ cm is the distance between the earth and moon, and $v_e = 4.62 \times 10^4$ cm/sec is the surface speed of the earth's equator. This amount of gravitational energy could yield energy at the same rate that solar energy is delivered to the earth's surface for a period of 1.5 million years. While large pieces of the moon might be economically torn loose from the moon by nuclear explosions so as to land on the earth, their impact on the earth would produce unfortunate seismic side effects, making the scheme impractical despite the large quantity of energy available.

Currently, energy is being extracted from the earth-moon gravitational system by the dissipation of energy as tidal friction. Only a small fraction is harnessed by the tidal power plant at Rance, France.

In order to absorb 1 percent of the sunlight from the sun using spaceships revolving around the sun (returning to the example above), it might take about $10^8$ gm of matter per person for a total of about $3.5 \times 10^{27}$ gm, or about the earth's mass. The gravitational energy that must be supplied to dismantle the earth is about $2.25 \times 10^{39}$ ergs (*see* Table IV), which could, in principle, be more than recovered by lowering to inside Mercury's orbit. It may also be noted that in sixty-six days this amount of energy is provided in 1 percent of the sun's radiation. Nuclear explosions should make the cost of dismantling small uninhabited planets energetically reasonable.

The gravitational energy in a spherical mass distribution of uniform density and radius R is given by

$$U = - G \int_0^M (m/r)\, dm = -3G\, M^2/5R \,,$$

(V.22)

where $m = 4\,\pi\,\rho\,r^3/3$, r being the radial variable and $\rho$ the density. This amount of energy has to be lost to form a planet (see Table IV).

INFALL ENERGY. Gas from interstellar space is trapped by the gravitational field of the sun. As this gas falls into the sun it gains kinetic energy and attains temperatures of the order of millions of degrees in the sun's corona. The gas is very tenuous, however, and no significant amount of energy is involved.

SOLAR WIND. The sun expels ions with high velocities which constitute the *solar wind*. Periodically the flux can be large enough to distort the earth's magnetic field to create magnetic storms. Spaceships traveling in the solar system could, in principle, utilize the solar wind for propulsion. The sails could be extended magnetic fields which would deflect the charged particles. The fields could be maintained by superconducting currents, so that no energy would be required. The superconducting state could be maintained by shielding the superconductors from the sun's rays.

The ionic wind might also be used to drive a magnetic windmill. The practicality of such devices depends upon the amount of energy available, which does not appear to be sufficient.

COSMIC RAYS. Although individual cosmic ray particles can be extremely energetic, the number of such particles is very small. The total energy flux is no more than starlight.

LIGHT PRESSURE. Photons from the sun exert a pressure on any surface upon which they are either absorbed or reflected. If the surface is moved work is done. The actual pressure exerted by sunlight turns out to be much too small to be of any practical significance.

MAGNETIC FIELDS IN SPACE. The total energy stored in the magnetic fields of the sun, earth and possibly Jupiter is not large, and there is no known way to extract such energy.

PHOTOELECTRIC EFFECT. Because a vacuum is readily available in space, sunlight could be used to eject photoelectrons from large surfaces, and by using grids and collectors useful electric power could in principle be generated. While the efficiency of such a process might not be high, it might be convenient.

ENERGY ALREADY DISSIPATED IN THE SOLAR SYSTEM. An insight into the energy balance in the solar system may be gained by considering the energy that has already been expended. In order for the sun to condense from interstellar matter to its present density of 1.41 gm/cm$^3$, $2.24 \times 10^{48}$ ergs had to be dissipated, Equation V.22. Similarly for the planets.

If it is assumed that the sun has been radiating at the present rate for 4.5 billion years since the formation of the planets, a total of $4.94 \times 10^{50}$ ergs of radiant energy have been dissipated. The total amount dissipated by the planets, assuming the same albedos as today, is about $1.14 \times 10^{42}$ ergs.

In order for the planets to be formed at their observed distances from the sun, energy equal to half their gravitational energy with respect to the sun had to be dissipated, about $1.97 \times 10^{42}$ ergs.

PARTITION OF ENERGY IN THE SOLAR SYSTEM. Although most of the energy in the solar system is not available to humans, a survey of all of the stores of energy helps to put things in perspective, and perhaps someday one may become of practical significance. Results are summarized in Tables IV and V.

The total energy contained in the solar system is $Mc^2 = 1.79 \times 10^{54}$ ergs where the total rest mass is $M = 1.99 \times 10^{33}$ gm, the kinetic and potential energies being insignificant.

If all of the matter in the solar system were allowed to condense down to the density of the atomic nucleus, $\rho_o = 9 \times 10^{13}$ gm/cm³, then the maximum gravitational energy that could be extracted is $4.22 \times 10^{52}$ from Equation V.22.

The thermal energy in the sun may be estimated from the expression for the internal energy for a monatomic ideal gas, $3nRT/2 = 1.91 \times 10^{48}$ ergs, where $R = 8.317 \times 10^7$ ergs/mole ° K , n is the number of moles given by the mass of the sun divided by the mean atomic mass of 1.3 gm, and the internal temperature is $10^{7°}$ K.

The entire solar system is in motion about the center of our galaxy with a velocity of $v = 2.25 \times 10^7$ cm/sec for a kinetic energy $Mv^2/2 = 5.03 \times 10^{47}$ ergs.

The radiant energy trapped inside the sun is given by $bT^4V = 1.08 \times 10^{47}$ ergs where $V = 1.41 \times 10^{33}$ cm³ is the volume of the sun and $b = 7.65 \times 10^{-15}$ erg/cm³ ° K⁴ is a universal constant.

If all of the planets were lowered to the surface of the sun, the gravitational potential energy of the solar system would be less by the amount

$$GM_s \sum_i m_i(1/r_s - 1/r_i) = 5.03 \times 10^{45} \text{ ergs} ,$$

where $M_s$ is the mass of the sun, $r_s$ is the radius of the sun, $r_i$ is the distance of the $i$th planet from the sun, and $m_i$ is the mass if the $i$th planet.

Assuming the sun is rotating as a solid sphere of uniform density, the kinetic energy is $I \omega^2/2 = 1.68 \times 10^{43}$ ergs where $I = 2M_s r_s^2/5 = 3.85 \times 10^{54}$ gm cm$^2$ is the moment of inertia and $\omega$ is the angular velocity of the sun.

TABLE IV.  ENERGY OF THE PLANETS

| Body | Gravitational Energy if Lowered to the Surface of Sun $10^{42}$ ergs | Kinetic Energy Revolving About Sun $10^{39}$ ergs | Rotation $10^{39}$ ergs | Gravitational Energy Lost to Form Planet $10^{39}$ ergs | Solar Energy Dissipated During Last $4.5 \times 10^9$ Yrs $10^{39}$ ergs |
|---|---|---|---|---|---|
| Mercury | 0.618 | 3.77 | 0.0000 | 0.0168 | 227 |
| Venus | 9.13 | 29.7 | 0.0000 | 1.48 | 173 |
| Earth | 11.4 | 26.6 | 0.0026 | 2.25 | 146 |
| Moon | 0.136 | 0.326 | 0.0000 | 0.00124 | 15.5 |
| Mars | 1.21 | 1.86 | 0.0000 | 0.0474 | 23.8 |
| Jupiter | 3580 | 1610 | 588 | 19750 | 465 |
| Saturn | 1070 | 262 | 124 | 2100 | 81.9 |
| Uranus | 164 | 19.9 | 3.22 | 111 | 3.97 |
| Neptune | 190 | 14.7 | 1.49 | 161 | 1.02 |
| Pluto | 1.14 | 0.134 | 0.0000 | 0.0478 | 0.118 |

TABLE V.  ENERGY IN THE SOLAR SYSTEM IN ERGS

| | |
|---|---|
| Total internal energy of solar system (rest mass energy) | $1.79 \times 10^{54}$ |
| Gravitational energy released if solar system were to condense to nuclear density | $4.22 \times 10^{52}$ |
| Thermal energy in the sun | $1.91 \times 10^{48}$ |
| Translational kinetic energy of solar system moving in the galaxy | $5.03 \times 10^{47}$ |
| Radiant energy in the sun | $1.08 \times 10^{47}$ |
| Gravitational energy released if planets lowered to sun | $5.03 \times 10^{45}$ |
| Rotational energy of the sun | $1.68 \times 10^{43}$ |
| Translational energy of planets about the sun | $1.97 \times 10^{42}$ |
| Rotational energy of the planets | $7.16 \times 10^{41}$ |

ENERGY DISSIPATED IN THE SOLAR SYSTEM
LAST 4.5 BILLION YEARS IN ERGS

| | |
|---|---|
| Radiant energy dissipated by the sun assuming the present rate of radiation | $-4.94 \times 10^{50}$ |
| Gravitational energy dissipated to form the sun | $-2.24 \times 10^{48}$ |
| Gravitational energy dissipated to form the planets | $-2.21 \times 10^{43}$ |
| Gravitational energy dissipated to place planets in their orbits about the sun | $-1.97 \times 10^{42}$ |
| Solar energy dissipated by the planets | $-1.14 \times 10^{42}$ |

# CHAPTER VI

# POPULATION AND CARBON IN THE BIOSPHERE

THE EXTREMELY LIMITED SUPPLY OF CARBON in the biosphere (or more precisely, the ecosphere) makes carbon-based life very sensitive to changes in the carbon cycle. Considering the large amounts of fossil carbon that are being currently dumped into the ecosphere, the question arises: Can a pertubation of the carbon cycle produce runaway oscillations leading to a depletion of carbon-based life?

## Carbon Mass as a Measure of Population

The application of thermodynamics to the study of the ecology of carbon-based life requires a measure of population which can be properly made quantitative in terms of physical principles. While a count of the number of individuals of a particular species yields a determinable number, it does not yield a number which can be readily used for physical analysis. Thermodynamically the number of individuals does not appear to be particularly significant. It would appear, to a first approximation, that the total mass of all the individuals of a particular species (perhaps per unit area) would be appropriate. Unfortunately, the mass of many organisms varies radically due to extraneous reasons. For example, birds, which must fly, conserve body weight by maintaining a low water content. Similarly, desert animals tend to weigh less by virtue of the necessity to conserve water. This problem might be circumvented by taking the measure as the total dessicated mass of the population; but this would fail to take into account the mass of bones, shells and such nonliving material which should perhaps not be included in the *population*. The total mass of some vital compounds might be appropriate, such as proteins. An easier measure would be the mass of an element that is associated with carbon-based life, such as carbon, oxygen or nitrogen. There appears to be no definitive way to obtain an appropriate measure of population at the present time. However, to a very reasonable first approximation it is suggested here that the total mass of carbon contained in a species may be used as a measure of the population, in short the *carbon population*.

The mass of carbon, or carbon population, has some fortunate features: All compounds associated with ordinary carbon-based life contain carbon. Carbon is the one element, contained quantitatively in all living carbon-based organisms, which is in very short supply in the biosphere and which may, therefore, be used to help distinguish between the living and nonliving. Populations of grossly dissimilar species may be readily placed on the same scale. For example, a population of 2 billion ($2 \times 10^9$) individual ants, having a mass of one milligram apiece, has a carbon population of just one elephant of 2000 kilograms mass (assuming the same proportion of carbon per gross weight for both species).

### Carbon Flux

While the standing biomass of a species, as measured by the carbon population, is significant, the time rate of flow of carbon into and out of the species is equally important. For example, the chemical energy made available to a species is ordinarily directly proportional to the carbon content of the food ingested; the carbon, being in essentially a reduced form, is oxidized to yield the energy. The chemical energy made available to other species by the species in question is proportional to the time rate of flux of carbon in the substances passing through the species in question. The time rate of flow of carbon through a species is proportional to the carbon population of the species times the mean metabolic rate of the species. While it is true that the carbon flux is proportional to the carbon population of a species, it should be noted that the metabolic rate may vary considerably. In particular, the metabolic rate of a species generally varies as about the reciprocal third power of the mass of a single individual (Odum, 1959). In the above example, the ants will account for a carbon flux roughly one thousand times greater than that through the single elephant.

### Iron Mass as a Measure of the Machine Population

Since machines also play a basic and inevitably more important role with time in the ecosphere, it is useful to indicate the measure of the population of machines which might be used which is equivalent physically to the carbon population of carbon-based life. If it is assumed that the quantity of machine life can be measured in terms of the iron content (although in 50 years the aluminum content may become more appropriate), then machine life and carbon life can be placed on the same scale by assuming that one atom of iron is equivalent to three-fourths atom of carbon, since three-fourths atom of carbon may be used to reduce one atom of iron from its iron oxide ore,

$$2Fe_2O_3 + 3C \rightarrow 3CO_2 + 4Fe . \qquad (VI.1)$$

In terms of mass, the atomic mass of iron is 55.8 gm/mole, while that of carbon is 12.0 gm/mole, so that the mass of machine life should be reduced by a factor of one sixth in order to place it on the same scale with carbon-based life. For example, an automobile with a mass of 1000 kgm is roughly equivalent to eight men of carbon mass of 20 kg each.

### The Limited Availability of Carbon

Considering the carbon population of ordinary carbon-based life in the biosphere before the advent of machines, one of the most universally limiting resources has always been carbon. Carbon life on the earth has always depended upon the available carbon contained in the atmosphere and oceans as carbon dioxide. It is only this carbon that is available for autotrophic organisms. This carbon is in extremely short supply. One of the basic population problems for the biosphere then becomes the determination of the biomass that can be supported with such a low concentration of available carbon.

The earth's atmosphere consists of the following primary constituents:

| Compound | Concentration in percent by mass |
|---|---|
| $N_2$ | 77.540 |
| $O_2$ | 20.800 |
| $H_2O$ | 0.700  variable |
| A | 0.927 |
| $CO_2$ | 0.033 |
| total | 100.000 |

It is clear that $CO_2$ is present only in a trace amount, water vapor generally exceeding the $CO_2$ concentration by a factor of about 20 and the rare gas argon exceeding it by a factor of 30. Since the atmospheric pressure is about $10^6$ dyne/cm² and the acceleration of gravity is about 980 cm/sec², the mass of carbon in the atmosphere above one square centimeter of earth is only 0.09 gm. It may be estimated that the oceans and lakes of the world have about 20 to 25 times as much dissolved $CO_2$ as contained in the atmosphere. When this source is included it still leaves only about 2 gm/cm². If the $CO_2$ from the atmosphere and oceans were all converted to carbohydrates (formula $CH_2O$), or to living material, only about 5 gm/cm² dry weight of living material would be obtained. This may be compared with a yield of 0.1 gm/cm² dry weight for a good wheat field in the midwestern part of the United States after only one brief three-month growing season.

Since under certain favorable circumstances, such as an unlimited supply of food, a species will increase in population exponentially with time (*see* e.g. Lotka, 1924a) in agreement with the Malthusian principle and observations, it is conceivable that within a very short time (as measured on a geologic time scale) all of the limited supply of available carbon might become tied up in just a few species. The fact that the amount of carbon available in the biosphere is extremely limited must be regarded as a very significant fact.

## Population of the Biosphere with Autotrophs Alone

The present hypothetical problem of a biosphere with autotrophs alone is presented as the first step in a series of progressively more difficult, but progressively more realistic, problems. With autotrophs only in the biosphere it may be assumed that carbon will only be extracted from the atmosphere and that the supply of $CO_2$ in the atmosphere and in the oceans will eventually become depleted. For a comparable mass of $CO_2$ and autotrophs in the biosphere it may be assumed that the rate at which autotrophs fixate carbon is proportional to the concentration of $CO_2$ in the atmosphere times the mass of autotrophs. However, for large concentrations of $CO_2$ and small concentrations of autotrophs, it may be assumed that the rate of fixation of carbon is proportional to the mass of autotrophs only. These ideas may be approximated mathematically by the equation

$$dL/dt = aCL/(1 + bC) \,, \qquad (VI.2)$$

where L represents the carbon mass of autotrophs in the biosphere, t the time, and C the mass of available carbon as $CO_2$ in the atmosphere and in the oceans, and a and b constants which depend upon a variety of factors such as rainfall, sunlight, minerals available and temperature. The fractional rate of increase in the carbon mass of autotrophs with time as a function of the mass of carbon, C, in the atmosphere is sketched qualitatively in Figure 9. For large concentrations, C, the fractional rate of increase in the carbon mass of autotrophs dL/at becomes constant at the value a/b. For concentrations that are low the fractional rate becomes proportional to aC.

If the total mass of carbon in the biophere, $C_o$, remains constant then C may be written as

$$C = C_o - L \,. \qquad (VI.3)$$

Substituting Equation VI.3 into VI.2 yields the differential equation

$$dL/dt = aL(C_o - L)/(1 + bC_o - bL) \,. \qquad (VI.4)$$

This may be integrated by quadratures yielding

$$\left(\frac{C_o - L_o}{C_o - L}\right)\left(\frac{L}{L_o}\right)^{1 + bC_o} = e^{C_o at} \,, \qquad (VI.5)$$

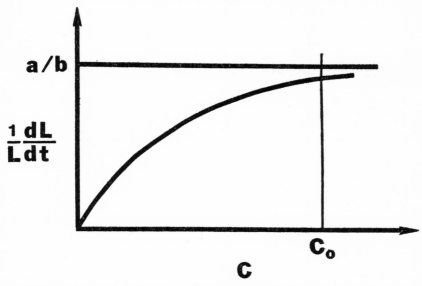

Figure 9. The fractional rate of increase of carbon mass of autotrophs in the biosphere as a function of the carbon content of the atmosphere, Equation VI.2, assuming a biosphere with autotrophs only that do not die.

where $L_o$ is a constant of integration representing the mass of autotrophs at time $t = 0$. This function, $L = L(t)$, is shown in Figure 10 along with C, as given by Equations VI.5 and VI.3. This result implies that after an infinite time the increase in autotrophs ceases (i.e. $dL/dt = 0$) when the $CO_2$ content of the atmosphere becomes exhausted (i.e. when $L = C_o$).

The general behavior of L with time may be indicated more simply by neglecting b. In this case

$$L \approx C_o/(1 + B_o e^{-C_o at}), \qquad (VI.6)$$

where $B_o$ is a constant equal to $(C_o - L_o)/L_o$.

This mathematical model might also be expected to approximate the case of crystals growing in a finite container from a supersaturated solution of salt before saturation is reached; or it might give the population of amoebas growing in a glass with a finite amount of nutrient prior to any deaths. It probably reflects to some degree of accuracy the situation on the surface of the earth almost 4.5 billion years ago when carbon fixating

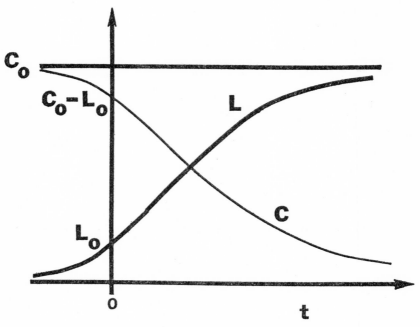

Figure 10. Variation with time, t, of autotrophic carbon, L, for a biosphere with autotrophs only that do not die. The thinner line suggests the variation of atmospheric carbon, C, with time.

processes were probably first becoming established. The original supply of available carbon on the earth was probably greater than it is today. This is suggested by not only the sizable deposits of fossil carbon and limestone, but also by the large quantities of carbon elsewhere in the solar system such as on Venus where the atmosphere is about 80 to 90 percent $CO_2$, on Mars where the atmosphere may be as much as 20 percent $CO_2$ and in the sun where carbon may represent almost 1 percent of the sun's mass. The large original supply of carbon on the earth may have been used up after only a few millions of years. Subsequently, life on the earth has apparently had to depend upon a continual return of carbon to the atmosphere.

### Population of the Biosphere with Autotrophs that Die

A hypothetical problem which is slightly more realistic than the example in the previous section is the problem of ascertaining the population of the biosphere if it consists of autotrophs that die. Autotrophs that die are assumed to lose their ability to fixate carbon from atmospheric $CO_2$. The dead bodies of autotrophs are assumed to accumulate with time and to remain undecomposed. The time rate of increase in the population of autotrophs is then assumed to be given by

$$dL/dt = aCL/(1 + bC) - akL , \qquad (VI.7)$$

where ak is a constant representing the fractional death rate for the autotrophs.

The rate of decrease of available carbon is given from Equations VI.2 and VI.3 by

$$dC/dt = - aCL/(1 + bC) . \qquad (VI.8)$$

Dividing Equation VI.7 by VI.8 and integrating yields the phase path

$$L = (1 - bk)(C_o - C) - k \ln(C_o/C) , \qquad (VI.9)$$

where the constant of integration has been chosen so that initially when $L = 0$, $C = C_o$, the total amount of carbon for the biosphere. This result is shown in Figure 11.

Substituting Equation VI.9 into VI.8 yields the expression

$$dC/dt = - aC[(1 - bk)(C_o - C) - k \ln(C_o/C)]/(1 + bC) , \qquad (VI.10)$$

which may be integrated numerically to yield C as a function of time as indicated in Figure 12.

From Equation VI.9 it may be noted that the available carbon in the atmosphere never goes to zero, since the autotrophs are all dead when $C = C'$ where $C'$ satisfies

$$(1 - bk)(C_o - C') - k \ln(C_o/C') = 0 . \qquad (VI.11)$$

The maximum population of autotrophs occurs for $C_m = k/(1 - bk)$, since the right side of Equation VI.7 is zero for this value and from Equation VI.9 the maximum population becomes

$$L_m = C_o(1 - bk) - k - k \ln C_o(1 - bk)/k . \qquad (VI.12)$$

If Equation VI.10 is used to find C as a function of time, then L may be obtained as a function of time from Equation VI.9.

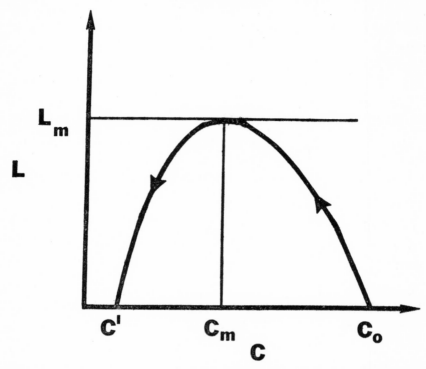

Figure 11. A curve indicating the relationship between the carbon mass of autotrophs, L, which die and the atmospheric carbon, C, Equation VI.9, assuming no carbon returned to the atmosphere from detritus.

For early times the second term in Equation VI.7 may be neglected, yielding the result of the previous section for early times. In the neighborhood of the maximum value for the population of autotrophs L may be replaced by a constant, Equation VI.12. Then Equation VI.8 may be integrated directly yielding the approximation

$$- \ln(C/C_m) + b(C_m{}^2 - C^2) = aL_m(t - t_m) , \tag{VI.13}$$

where $t_m$ is the time at which the maximum population occurs. Neglecting b, the mass of atmospheric carbon is approximately given by

$$C = C_m \exp\left[-aL_m(t - t_m)\right] , \tag{VI.14}$$

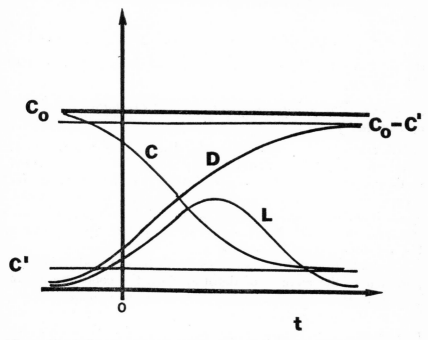

Figure 12. A diagram indicating the carbon mass of autotrophs, L, detritus, D, and atmospheric carbon, C, as functions of time, t, for the hypothetical case of no carbon being returned to the atmosphere.

in the neighborhood of $t = t_m$. Finally after a long period of time C becomes small so bC and $C/C_o$ may be neglected compared with unity in Equation VI.10, yielding

$$C = C_o \exp(e^{-akt} - C_o/C_m) . \qquad (VI.15)$$

For $t \to \infty$ the final value for the carbon left in the atmosphere is given by

$$C' = C_o \exp(- C_o/C_m) , \qquad (VI.16)$$

in agreement with Equation VI.11 if $C'$ is neglected compared with $C_o$.

The amount of dead autotrophs or detritus, D, may be obtained from C as a function of time and Equation VI.9 by noting that the total carbon in the ecosphere, including the mass of detritus, must remain constant at the value

$$C_o = C + L + D , \qquad (VI.17)$$

or

$$D = bk(C_o - C) + k \ln(C_o/C) , \qquad (VI.18)$$

the final value being $D' = C_o - C'$. These results are also indicated in Figure 12.

### Biosphere with Autotrophs, Detritus and Oxidation of Detritus

Here it will be assumed that the dead bodies of autotrophs, or detritus, is oxidized, thereby permitting carbon to be returned to the atmosphere. This mathematical model, thus, provides for a cyclic flux of carbon and, therefore, provides a more realistic approximation of the actual ecosphere. This model not only provides for the possibility of steady states, but it also introduces the interesting possibility of oscillations.

The rate that detritus is oxidized will be assumed to be proportional to the amount of detritus, since the concentration of oxygen in the atmosphere may be assumed to remain constant. The precise mechanisms whereby detritus is oxidized need not be explored here. It may be noted, however, that fires, slow oxidation, detritus feeders and apparently volcanoes all play a role. Since it was probably only during early geological times that there were large concentrations of $CO_2$ in the atmosphere, there is no need here to consider the bC appearing in Equations VI.2 and VI.7 which may be neglected compared with unity. Autotrophs themselves also return carbon to the atmosphere in proportion to the population of autotrophs during plant respiration. Including this oxidation rate as aeL, where e is a new constant, Equations VI.7, VI.8 and VI.17 yield the appropriate set of equations:

$$dL/dt = aCL - akL - aeL ,$$
$$dC/dt = ahD - aCL + aeL ,$$
$$dD/dt = akL - ahD ,$$
$$C_o = C + L + D ,$$

$$(VI.19)$$

where ah is a constant giving the fractional rate of oxidation of detritus. Figure 13 indicates the flow of carbon in the ecosphere as given by Equation VI.19.

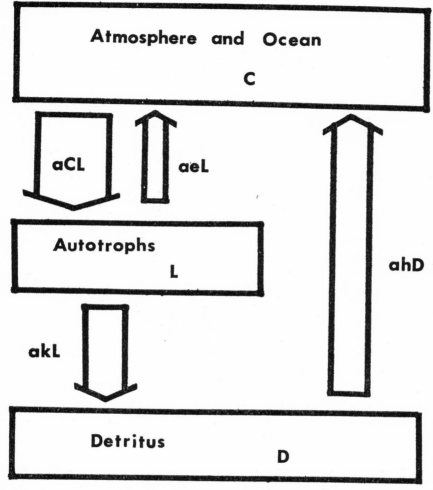

Figure 13. A diagram of the cyclic flow of carbon in the ecosphere assuming the existence of autotrophs, detritus and the oxidation of detritus (not to scale).

Using the last of Equation VI.19 the population of detritus may be eliminated to yield the two-population problem:

$$dL/dt = aCL - a(k + e)L,$$
$$dC/dt = ah(C_o - C) + a(e - h)L - aCL .$$

$$(VI.20)$$

At equilibrium the time rates of change go to zero; thus, the steady state values become

$$\underline{C} = k + e \,,$$
$$\underline{L} = h(C_o - e - k)/(h + k) \,,$$
$$\underline{D} = k(C_o - e - k)/(h + k) \,.$$

$$(VI.21)$$

NUMERICAL VALUES OF THE CONSTANTS. It may be assumed that today the biosphere is essentially in steady state equilibrium, so that the values of $\underline{C}$, $\underline{L}$ and $\underline{D}$ can be estimated from observations. Numerical estimation of the carbon content in the various reservoirs on the earth and the average rate of flow between these reservoirs are quite uncertain (e.g. Welte, 1969; Bender, 1972). Only the carbon content of the atmosphere is known to a reasonable accuracy (2 places) and is $0.64 \times 10^{18}$ gm. Not all of the carbon in the ocean is immediately exchangeable with the atmosphere and is readily available for photosynthesis. Assuming thirty times more carbon in the ocean is available for photosynthesis than in the atmosphere, the atmosphere-ocean reservoir may be roughly estimated to contain $20 \times 10^{18}$ gm of carbon. If the mass of cycling detritus is roughly estimated to be 4 percent of this amount and if the mass of autotrophs is roughly taken to be 0.6 percent of this amount, then the following approximate estimates are obtained:

$$\underline{C} = 20\,, \qquad \underline{L} = 0.1\,, \qquad \underline{D} = 0.8\,, \qquad (VI.22)$$

in units of $10^{18}$ grams.

If it is roughly estimated that autotrophs fixate $8 \times 10^{-5}$ gm of carbon per day per square centimeter (for a 24-hour day, 365 days a year) averaged over the entire earth, then the whole biosphere fixates about $4 \times 10^{14}$ gm of carbon per day. If for every carbon atom fixated another one-third carbon atom on the average is used in respiration, then $a\underline{CL} = 6 \times 10^{14}$ gm/day and $ae\underline{L} = 2 \times 10^{14}$ gm/day. From these estimates and Equation VI.22

$$a = 3.0 \times 10^{-22} \text{ gm}^{-1} \text{day}^{-1}\,; \qquad (VI.23)$$

and from the last of Equations VI.19 and VI.21 the remaining constants in Equations VI. 20 are given by

$$C_0 = 20.9, \quad e = 6.7, \quad k = 13.3, \quad h = 1.7,$$
$$(VI.24)$$

in units of $10^{18}$ gm. While these numerical values are very uncertain, it is hoped that they may provide an order of magnitude estimate.

SOLUTION NEAR STEADY STATE EQUILIBRIUM. Near steady state equilibrium C and L appearing in Equations VI.20 may be expanded about their equilibrium values, $\underline{C}$ and $\underline{L}$. Thus letting

$$x = C - \underline{C}, \qquad y = L - \underline{L}, \qquad (VI.25)$$

the appropriate set of equations from Equations VI. 20 becomes

$$dy/d\tau = xy + \underline{L}x,$$
$$dx/d\tau = -xy - (h + \underline{L})x - (h + k)y,$$
$$(VI.26)$$

where

$$\tau = at. \qquad (VI.27)$$

Near steady state equilibrium second powers in x and y may be neglected (the first terms on the right of Equations VI.26 being neglected), which yields the differential equation of second order for either x or y

$$d^2y/d\tau^2 + (h + L) \, dy/d\tau + (h + k)L \, y = 0. \qquad (VI.28)$$

The solution to this equation is

$$y = y_0 e^{-t/t'} \sin(2\pi t/T + \phi), \qquad (VI.29)$$

where $y_0$ and $\phi$ are constants of integration. This solution yields a rapidly damped solution (on a geological time scale with the present numerical estimates) which reduces to $1/e$th the original value in only

$$t' = 2/a(h + \underline{L}) = 10 \text{ years}; \qquad (VI.30)$$

it has a periodic component with a period of

$$T = (4\pi/a)[4k\underline{L} - (h - \underline{L})^2]^{-\frac{1}{2}} = 69 \text{ years}.$$

$$(VI.31)$$

Since the period is greater than the decay time $t'$, the periodicity would not be noticeable. By virtue of the very short time constant of only 10 years the biosphere is evidently in a very stable steady state equilibrium with regard to cycling carbon. There need be no fear of runaway oscillations. The slow withdrawal of carbon from the ecosphere becomes deposited as fossils, limestone, or escapes from the earth has not been considered here. At the present time it may be assumed that permanent withdrawals of carbon are proceeding at a very slow rate with a time constant of the order of $10^9$ years.

For comparison with material to be presented in the following Chapter a phase diagram relating L and C equations VI.20, is sketched in Figure 14. The curves on which $dL/dC$ is zero or infinite, as obtained by dividing Equations VI.20, are shown. The phase trajectory is indicated.

### Biosphere with Autotrophs, Detritus and Heterotrophs

Since man is a heterotroph and much of his preoccupation is with other heterotrophs, the present section explores the possible effect of heterotrophs on the biosphere.

To a good first approximation it will be assumed that the class of all heterotrophs does not directly affect the population of autotrophs. While it is easy to find numerous individual cases where heterotrophs kill the autotrophs upon which they feed and apparently do effect the population of such autotrophs directly, there are still many more cases where the heterotrophs have no such effect on the autotrophs upon which they feed. For example, grazing animals in a temperate zone may eat live grass for only four months of the year; the rest of the year they must eat dead grass. Thus, their role must be counted as at least two-thirds detritus feeders. Actually their effect on the live grass is ordinarily small, so that such grazing animals should ordinarily be regarded as almost 100 percent detritus feeders. Worms that eat apples do not prevent the tree from flourishing. Except for a slight effect on the reproduction of apple trees, such worms

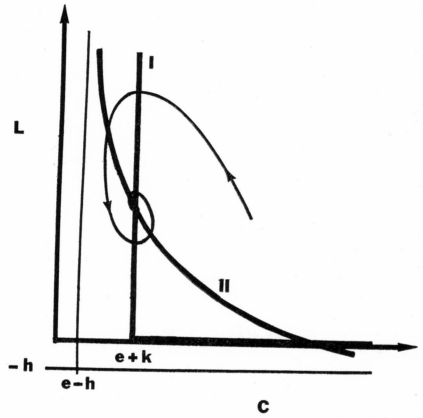

Figure 14. Phase diagram for Equations VI.20 indicating the characteristic curves I on which $dL/dC$ is zero and II on which $dL/dC \to \infty$, and indicating the phase trajectory.

may be regarded as detritus feeders. Tent caterpillars may strip all of the leaves off a large tree, but this occurs in late summer just before the tree becomes essentially dormant, and the tree puts out its full complement of leaves the next year, showing no ill effects. Tent caterpillars are, thus, essentially detritus feeders.

In addition to citing examples, there is ample theoretical justification for considering the class of all heterotrophs as detritus feeders to a good first approximation. In particular, in order for an ecosystem to evolve toward a time average maximum biomass it must evolve a system which maximizes the wel-

fare of its primary producers, or autotrophs. Material that flows
from the autotrophs to the heterotrophs must be material that
is no longer of any value to the class of all autotrophs. It may
also be noted that predators (heterotrophs) must evolve so as to
benefit their obligate prey (autotrophs), since the predators'
continued existence is necessarily dependent upon the prey's
survival. Given sufficient time the relationship should evolve to
a symbiotic relationship. Consequently, the class of all heterotrophs cannot be viewed as actively detrimental to the class of
all autotrophs; instead heterotrophs as a class should be viewed
as passively receiving the detrius from the class of autotrophs.

In the present analysis direct oxidation of detritus such as
occurs slowly and in forest and brush fires will be neglected as
compared with the oxidation achieved by heterotrophs. The
recent oxidation of fossil fuels by machines will also be neglected here. The role of volcanoes in returning $CO_2$ to the atmosphere is hard to judge. However, the earth has apparently under-
gone many very long periods in which there was little or no
mountain growing activity and therefore no volcanoes; yet no
depletion of $CO_2$ seems to have ever occurred. Consequently, it
will be assumed here that the oxidation of detritus occurs only
through heterotrophs. For values not far from steady state equilibrium the rate of consumption of detritus by heterotrophs may
be taken as proportional to the product of the population of
detritus and the population of heterotrophs, or abDM where M
is the mass of carbon in heterotrophs and ab is a constant.

If the rate that heterotrophs deliver carbon to the atmosphere
is taken as proportional to the population of heterotrophs, or
aiM where ai is a constant, and if the rate that heterotrophs die
(becoming detritus) is taken as proportional to the population
of heterotrophs, or ajM where aj is a constant, then the carbon
flow pattern is indicated in Figure 15 and is specified by the
following set of equations:

$$dC/dt = - aCL + aeL + aiM ,$$
$$dL/dt = aCL - akL - aeL ,$$
$$dD/dt = akL - abDM + ajM , \qquad (VI.32)$$
$$dM/dt = abDM - aiM - ajM$$
$$C_o = C + L + D + M .$$

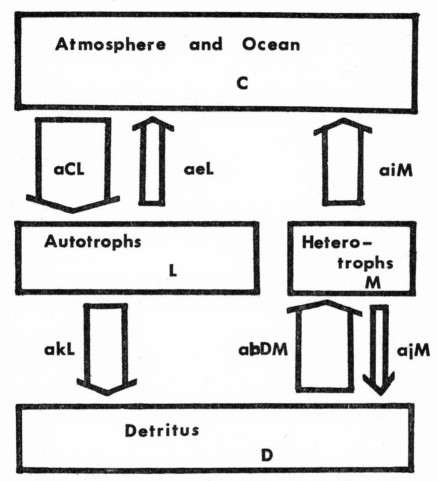

Figure 15. A diagram of the reservoirs and the flow of carbon in the ecosphere for the case of autotrophs and heterotrophs as detritus feeders. (The diagram is not to scale.)

NUMERICAL VALUES OF THE CONSTANTS. It may again be assumed that the biosphere is essentially in a steady state equilibrium, so that the time derivatives in Equations VI.32 may be set equal to zero and the constants may be estimated from observations. The amount of available carbon in the atmosphere and ocean, in autotrophs, and in detritus may be estimated as before, Equation VI.22. The rate of fixation of carbon by photosynthesis may be assumed as before, and a, e and k, using the second of Equations VI.32 for $dL/dt = 0$, yields the values given before in Equations VI.23 and VI.24. It will be estimated here that the mass of heterotrophs is roughly 10 percent of the mass of autotrophs (noting that in oceans the mass of heterotrophs relative to autotrophs is much greater than on land); thus, summarizing, using the last of Equations VI. 34,

$$C = 20, \quad L = 0.1, \quad D = 0.8, \quad M = 0.01,$$
$$\overline{C}_o = 20.91, \quad e = 6.7, \quad k = 13.3, \quad \text{(VI.33)}$$

in units of $10^{18}$ gm.

It is roughly estimated that 10 percent of the flow of carbon into heterotrophs is returned to detritus, then $bD = 10j$, and the remaining constants appearing in Equation VI.$\overline{32}$ have the estimated values:

$$i = 133 \times 10^{18} \text{ gm}, \quad j = 14.8 \times 10^{18} \text{ gm},$$
$$b = 185, \quad a = 3.0 \times 10^{-22} \text{ gm}^{-1} \text{ day}^{-1},$$
$$\text{(VI.34)}$$

where b is dimensionless.

SOLUTION NEAR STEADY STATE EQUILIBRIUM. Using the last of Equations VI.32 to eliminate the variable D, Equations VI.32 reduce to

$$dC/adt = - CL + eL + iM,$$
$$dL/adt = L(C - k - e), \quad \text{(VI.35)}$$
$$dM/adt = bM(C_o - C - L - M) - i - j.$$

Expanding the equations about the steady state equilibrium by letting

$$x = C - \underline{C}, \quad y = L - \underline{L}, \quad z = M - \underline{M},$$
$$\text{(VI.36)}$$

where

$$\underline{C} = k + e = 20,$$
$$\overline{L} = i[b(C_o - k - e) - i - j]/b(i + k) = 0.1,$$
$$\overline{M} = k[b(C_o - k - e) - i - j]/b(i + k) = 0.01,$$
$$\underline{D} = (i + j)b = 0.8, \quad \text{(VI.37)}$$

where the numerical estimates are in units of $10^{18}$ gm, Equations VI.35 become

$$dx/adt = -x(y+L) - ky + iz ,$$
$$dy/adt = x(y+L) ,$$
$$dz/adt = -b(x+y+z)(z+M) . \qquad (VI.38)$$

Neglecting second powers in the small variations from equilibrium, x, y and z, using the second of Equations VI.38 to eliminate the variable x, the equations reduce to

$$d^2y/d\eta^2 + dy/d\eta + ky/L - iz/L = 0 ,$$
$$dy/d\eta + y + (i/bk)dz/d\eta + z = 0 , \qquad (VI.39)$$

where the independent time variable has been chosen as the dimensionless quantity

$$\eta = a\underline{L}\, t . \qquad (VI.40)$$

The first of Equations VI.39 may be substituted in the second to eliminate the variable z, thus yielding the third order differential equation

$$y''' + \alpha\, y'' + \beta\, y' + \gamma\, y = 0 , \qquad (VI.41)$$

where prime denotes differentiation with respect to $\eta$ and the constants $\alpha$, $\beta$ and $\gamma$ are given by the dimensionless quantities

$$\alpha = 1 + bk/i = 19.5$$
$$\beta = k(1+b)/L + bk/i = 24,800$$
$$\gamma = bk(i+k)/\overline{i}\underline{L} = 27,100 . \qquad (VI.42)$$

Assuming an exponential solution of the form

$$y = A\, e^{-p\eta} \qquad (VI.43)$$

where A and p are constants and p must satisfy the cubic equation

$$p^3 - \alpha\, p^2 + \beta p - \gamma = 0 . \qquad (VI.44)$$

The solution to a cubic equation can be expressed in closed form, but it will be sufficient here to note that the roots have the approximate numerical values given by $p \approx \gamma/\beta$ and $p^2 - \alpha p + \beta \approx 0$; or

$$p_1 = 1.09 \,,$$
$$p_2 = 9.75 + 157 \text{ i} \,,$$
$$p_3 = 9.75 - 157 \text{ i} \,, \qquad\qquad (\text{VI}.45)$$

where $i = \sqrt{-1}$. The general solution of Equation VI.41 takes the form

$$y = A_1 e^{-t/t'} + A_2 e^{-t/t''} \sin(2\pi t/T + \phi) \,, \qquad (\text{VI}.46)$$

where $A_1$, $A_2$ and $\phi$ are constants of integration. From Equation VI.40 and the numerical values, the first of Equations VI.45, the decay constant for the nonoscillating component is

$$t' = 1/p_1 a\underline{L} = 84 \text{ years} \,. \qquad\qquad (\text{VI}.47)$$

The decay constant for the oscillating component may be similarly found from the real part of $p_2$ or $p_3$, Equation VI.45, or

$$t'' = 9.4 \ \text{ years} \,; \qquad\qquad (\text{VI}.48)$$

and the period of oscillation from the imaginary part of $p_2$ or $p_3$ becomes

$$T = 3.6 \text{ years} \,. \qquad\qquad (\text{VI}.49)$$

Both components of the solution, Equation VI.47, Equations VI.48 and VI.46, are rapidly damped on a geologic time scale, indicating a very stable carbon cycle when heterotrophs are also included. Again there need be no fear of any runaway oscillations. Since the period of oscillation for the second component of the solution, Equation VI.46, is about one-third the decay time, the oscillations might be observable. See Figure 16.

Before a time varying solution such as Equation VI.46 could be observed a displacement from steady state equilibrium would have to first occur. The ice ages indicate that such displacements from equilibrium have actually occurred in the past. Such perturbations might be caused by changes in the solar energy output, a changing greenhouse effect, or effects still unknown.

### Possible Observations of the Time Variation of the Carbon in the Atmosphere

It is known that the ratio of the atmospheric radioactive carbon 14 to ordinary carbon 12 has an erratic secular change with

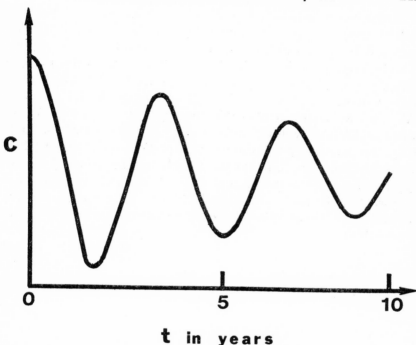

**t in years**

Figure 16. The damped oscillations of the carbon content of the atmosphere following a perturbation under the assumptions in the text, the second term on the right of Equation VI.47. The scale for C is arbitrary, as well as the origin.

time as observed in tree rings over the last 8000 years (Olsson, 1971). These observations are not spurious, since the same time variations are observed in trees around the world. Radioactive carbon is manufactured in the atmosphere from slow neutrons derived originally from cosmic rays and solar flares (Lingenfelter, 1963). While the observed variations might be explained in terms of solar activity, the solar wind shielding the earth or else solar flares injecting more neutrons, they may also in part be explained by the existence of damped oscillations as discussed above.

In particular, the set of equations governing the quantity of $^{14}C$ in the various carbon reservoirs on the earth may be taken as slight modifications of Equations VI.32. Including a constant rate of manufacture in the atmosphere and a constant fractional radioactive decay rate in each of the reservoirs, the equations may be appropriately modified. In addition, due to the isotopic effect, the coupling coefficients governing the flow between reservoirs will be slightly altered. As a consequence, the damping rate for radioactive carbon fluctuations (*cf*. Equations VI.30, VI.48 and VI.47) will differ from that of ordinary carbon. Similarly, the periodicity will differ (*cf*. Equations VI.31 and VI.49). The net effect may, thus, explain the observed apparently erratic secular variations in the ratio of $^{14}C/^{12}C$, assuming an occasional perturbation in the rate of production of $^{14}C$. No precise numerical estimates are possible because of the uncertainties in the observations.

# CHAPTER VII

# POPULATION OF AN INDIVIDUAL SPECIES

IN THE PREVIOUS CHAPTER it was possible to get a rough estimate of the stability of the whole biosphere as far as the
carbon cycle was concerned by considering small displacements
from steady state equilibrium of the mass of carbon in autotrophs, detritus, heterotrophs, and in the atmosphere and oceans.
Mathematically the situation did not prove to be too difficult. In
the present Chapter, however, the much more interesting question of the survival of a single species leads to greater mathematical difficulties, because small displacements from steady state
equilibrium can no longer be assumed.

Perhaps the present mathematical investigation of populations properly belongs to mathematical ecology (e.g. Pielou,
1969) rather than to ecophysics; but, since the boundary between theoretical physics and mathematics is not always precise
and since predicting populations is a central problem in ecology,
it was decided to include this material here.

There has been a little work done over the years on the
mathematical problem of populations (e.g. see the excellent review article by Goel et al., 1971; the multiple-authored two volumes edited by Patten, 1972; Slobodkin, 1963). There have also
been a few attempts to fit theories to experimental and field
data (e.g. Neyman et al., 1956; Leigh, 1969). Much of this work,
while interesting and mathematically instructive, is not par-

147

ticularly realistic or practical, because it fails to make realistic assumptions about the coupling between populations and other factors affecting populations.

While the statistical approach (which is certainly necessary for small populations) of Kerner (1971, 1957, 1959, 1961) and others is significant, research in the present Chapter is limited to large populations that can be presumably represented by continuous variables.

## Evolution and the Extinction of Species

Evolution depends upon the extinction or survival of particular species; consequently, one of the more important problems in ecology is predicting the survival or extinction of a species. Darwin (1859) viewed species as evolving morphological and behavioral features to gain advantages in food gathering and other activities related to the survival of an individual member of a species. While this view is undoubtedly correct to a good first approximation, it is not the individual organism that becomes extinct; it is the total population of a species that goes to zero. A species must evolve to satisfy not only the needs of its individual members, but it must also evolve morphological structures and patterns of behavior which allow the total population of the species to remain above zero.

The distinction between the survival of an individual member of a species and the survival of the entire population of a species can be made apparent by considering a few examples: The population of robins could not survive without the extreme effort of the mother and father robin during the crucial days just prior to and just after the baby robin leaves the nest. Yet this great effort by the individual parent does not contribute to the survival of the individual parent. In some species of spiders the male gives his life happily, it would appear, to impregnate a female to continue the species. Warrior ants readily allow themselves to be destroyed in the defense of their nest. The brilliant colors of an American cardinal must allow predators to discover more easily the location of the bird. The bright colors of American birds in general must mitigate against their individual survival. Similarly, the beautiful and attention-attracting songs of European birds must contribute to the predatory harvest of European birds. Yet such displays of color and song apparently contribute to the survival of the species by optimizing the apportionment of territory among individual members (or mated pairs) of the species (*see* Chapter X).

## Need for Mathematical Models to Predict Population

While it is possible to read many accounts in which an author purports to know the reason for the variation of the size of a population without a mathematical theory, a very brief examination of even the simplest situations using the simplest of mathematical models indicates the futility of such an attempt. Descriptions, analysis and explanation of the complicated results of field observations and experiment must necessarily rest upon the use of mathematics as a language. While numbers may test the correctness of a mathematical theory or model, the purpose of a mathematical model is actually to display the interrelationships between the variables involved in a meaningful manner.

In order to present the mathematics in an orderly manner, a number of simpler classical problems will be treated first.

## The Problem of Malthus

Malthus (1824) in the early eighteen hundreds predicted that a population of humans would increase geometrically until the population was such as to provide the minimal food supply per person. Malthus' theory has been attacked more often than supported; yet the essential feature that a population is more or less limited by the available food supply is undoubtedly correct. In addition, the fact that a small population in an infinite supply of food increases geometrically with time cannot be doubted. Subsequent sections will present more realistic models which still do not detract from the initially important observations of Malthus.

Mathematically Malthus' theory states that a population y reproduces at a rate proportional to the population y. Assuming a constant probability b of death per unit of time per individual (which is within the spirit of the Malthusian theory), the net rate of increase in population then becomes

$$dy/dt = ay - by = cy ,$$

(VII.1)

where a, b and $c = a + b$ are constants. Integrating this equation yields the time rate of increase (or decrease if c is negative) of the population

$$y = A\ e^{ct}, \tag{VII.2}$$

where A is a constant indicating the population at time $t = 0$. If food x is supplied at a constant rate,

$$dx/dt = e, \tag{VII.3}$$

and if the minimum rate that food must be supplied to maintain the life of an individual member of the population, on the average, is

$$(dx/dt)_0 = f, \tag{VII.4}$$

then the final population that can be supported according to Malthus is

$$y_m = e/f. \tag{VII.5}$$

From Equation VII.2 this population is reached in the finite time

$$t' = c^{-1}\ \ln(e/fA). \tag{VII.6}$$

These results, Equations VII.2, VII.5 and VII.6, are shown in Figure 17.

### The Problem of Malthus with a Density Dependent Death Rate

Early ideas also attempted to take into account the density effects of population, so that the discontinuity in the slope of y, which occurs in the theory of Malthus at $t = t'$, Figure 17, could be eliminated (Verhulst, 1844 and 1846; Pearl and Reed, 1920 and 1921). It was reasoned that the death rate due to communicable diseases was proportional to the rate of contact between those infected and capable of transmitting the disease and those not infected. It was assumed that such a frequency of contact was proportional to the contact between any two individuals in general or to $y^2$. This approximation is not rigorously correct; the diseased population and the well population must be treated

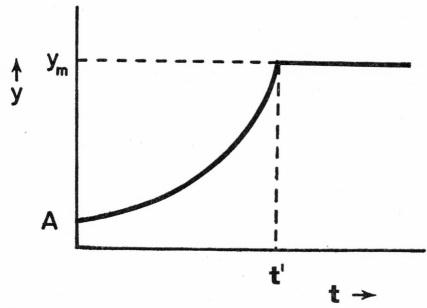

Figure 17. A curve indicating the variation of population y with time t starting with an excess food supply and an initial population A, according to Malthus, Equation VII.2.

as two interacting populations, which can lead to significantly different results from the simple assumption that the death rate is proportional to $y^2$. However, there appear to be some real density effects (evolutionarily evolved) which do vary as functions of the density of population. It appears that many species require a certain minimum area or volume per individual. If the area or volume per individual becomes less than the minimum, a complicated series of degenerative and population reducing processes come into play. Similarly, the rate of cannibalism, being based upon the chance meeting of an adult with a larva, in the tribolium beetle appears to depend upon the density of the population.

Such density effects do not seem to come into play very often in nature, and their importance has probably been overestimated, as mentioned by Andrewartha (1961a).

Assuming a density dependent death rate proportional to $y^2$, Equation VII.1 is generalized to

$$dy/dt = cy - gy^2 , \qquad\qquad (VII.7)$$

where c and g are constants. This equation may be immediately integrated, yielding the continuous logistic curve

$$y = (c/g)/(1 - ke^{-ct}) , \qquad\qquad (VII.8)$$

where k is a constant of integration. This result is indicated in Figure 18. The asymptomatic value of the population,

$$y_m = c/g , \qquad\qquad (VII.9)$$

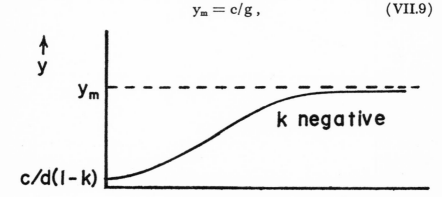

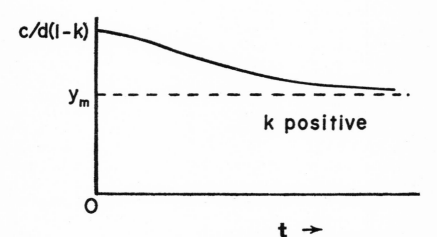

Figure 18. Population y as a function of time t for the Malthusian theory plus a density-dependent term proportional to $- gy^2$, Equation VII.8.

does not necessarily relate in any way to the rate at which food is supplied. If the present maximum value were the same as the previous case, then $c/g = e/f$; but such an equality would be fortuitous. In general a population appears to maintain itself somewhat below the maximum population that would be maintained in terms of the food supply alone, which would mean that in general $c/g < e/f$.

The curve, Equation VII.8, has been fitted to a number of experimental situations (Lotka, 1924b). While there appears to be some agreement in some cases, the density dependent effect is probably not correctly represented as simply proportional to $y^2$; so that only general overall qualitative behavior can be predicted by this simple theory.

### The Problem of Volterra

Another classical population problem which includes the interaction of two populations, predator and prey, was investigated by Volterra (1931, 1926, 1927) following Lotka. In this problem it is assumed that the predator (lynx, for example) eats the prey (arctic hare, for example) (Elton, 1927) at a rate that is proportional to the product of the number of predators y times the number of prey x.

Assuming that these populations refer to an area density, the rate that predators catch prey may be estimated as follows: If a predator can catch its prey when the prey is within a distance s of the predator, on the average, then the total effective area searched by a predator per unit time is 2sv where v is the mean speed of the predator (assuming stationary prey). The total area searched per unit time for the whole population of predators becomes 2svy. The frequency with which a prey is caught in this area is then given by multiplying this by the density of prey, or 2svxy. This result indicates the validity of assuming that the rate predators eat prey is proportional to the product xy. The same dependence of the rate of consumption occurs for many food supplies, for example, nuts sought by squirrels. However, it must be noted that this relationship holds only so long as the predator needs to search for food. If the food supply or the population of prey becomes sufficiently large, the rate at which food is eaten or prey caught will vary only with the number of predators, each predator being able to eat only a certain maximum amount. Volterra's condition of want where the rate at which prey are caught is proportional to xy appears to actually be rare in nature. More realistic assumptions will be explored in subsequent sections.

Volterra assumed that the rate at which the prey reproduce is proportional to the number of prey. It will be indicated later that this assumption is also usually not true, being valid only for very small populations. If the death rate of the prey, apart from those eaten by predators, is assumed to be proportional to the number of prey, then the net rate of increase in the prey population according to Volterra is given by

$$dx/dt = ax - bxy \, , \qquad (VII.10)$$

where a and b are constants.

Volterra further assumed that the reproduction rate of predators was proportional to the amount of food they eat or to xy. While this assumption appears to be reasonable at first, the consumption of food or the availability of food rarely has, in fact, such a direct effect upon the birth rate. This assumption also ignores many other determiners of the birth rate which are usually more important than the product xy. Finally Volterra assumed that the death rate of predators was proportional to the number of predators. Under these assumptions the net rate of increase in the number of predators becomes

$$dy/dt = cxy - ey \, , \qquad (VII.11)$$

where c and e are constants. Combining Equations VII.10 and VII.11 then gives the coupled set of differential equations:

$$dy/dt = cxy - ey \, ,$$
$$dx/dt = ax - bxy \, . \qquad (VII.12)$$

While it will be shown below that this set of equations cannot represent an actual predator-prey relationship (primarily due to the unrealistic assumption that the birth rate or reproduction rate of predators is proportional to xy), it is mathematically instructive to continue with this simple historic example.

Dividing the first of Equations VII.12 by the second and integrating yields

$$\frac{e^{by}}{(by)^a} \, \frac{e^{cx}}{(cx)^e} = k, \qquad (VII.13)$$

where k is a constant of integration. This is a closed curve (if k is chosen appropriately) in the x-y plane as shown in the first quadrant of Figure 19. This closed curve may be constructed by defining $\eta$ and $\xi$ such that

$$\eta = e^{by}/(by)^a, \qquad \xi = (cx)^e/e^{cx}. \qquad (VII.14)$$

Then Equation VII.13 becomes simply

$$\eta = k\,\xi, \qquad (VII.15)$$

as shown in the third quadrant of Figure 19. In the second quadrant $\eta$ is shown as a function of y and in the fourth quadrant $\xi$ is shown as a function of x. The graphical construction of the closed curve is then apparent. The slope k must not be less than $a^{-a} \, e^{-e} \exp(a + b)$ for real values of x and y.

As the closed curve in Figure 19 is traced out in time oscillations in the predator-prey populations occur in time. Volterra assumed that populations might actually be able to sustain such oscillations indefinitely. This would represent a stable, though oscillatory solution. The observed periodic fluctuations in the populations of lynxes and arctic rabbits (Elton, 1927) appear to support such a possibility. But one small realistic modification of Volterra's theory may be introduced to reveal the fact that Volterra's problem is actually unstable and always leads to the extinction of the predator.

In particular, let it be assumed that a certain finite number of predators $y_o$ and prey $x_o$ are necessary in order for reproduction in these species to occur. Perhaps $y_o$ and $x_o$ might be set equal to two for the proverbial mated pair. Since the reproduction rates go to zero when $x < x_o$ and $y < y_o$, Equations VII.12 are modified to read

$$dy/dt = cx(y - y_o) - ey,$$
$$dx/dt = a(x - x_o) - bxy. \qquad (VII.16)$$

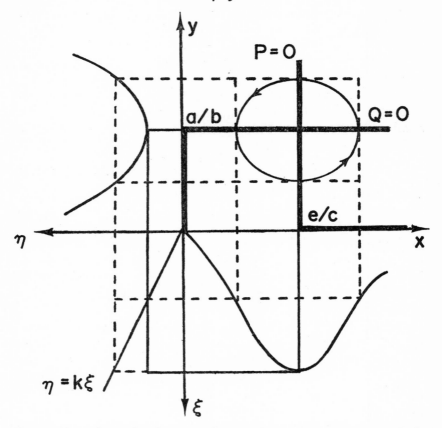

Figure 19. Relationship between the population of predators y and the population of prey x (first quadrant) according to Volterra. The remaining quadrants indicate how the phase trajectory may be constructed.

Before proceeding with the analysis of these equations, it will first be useful to develop some general mathematical tools.

### Phase Diagrams and the Behavior of the Phase Trajectory Near a Singularity

When considering the survival potential of a single species one is almost always led to a two-population problem; consequently, the two-population problem is one of the more important problems in ecology. A facility with the two-population problem and with phase diagrams can be useful. Two coupled populations yield a pair of differential equations that may be written in the general form

$$dy/dt = P(x,y),$$
$$dx/dt = Q(x,y). \qquad (VII.17)$$

A general discussion of such a pair of equations may be found in standard texts (e.g. Davis, 1962; Sansome, 1964). A phase trajectory (i.e. a solution $y = y(x)$ obtained by dividing Equations VII.17 and integrating) on the x-y plane will have zero slope when $P(x,y) = 0$ and will have infinite slope when $Q(x,y) = 0$. The intersection of the two characteristic curves,

$$P(x,y) = 0,$$
$$Q(x,y) = 0, \qquad (VII.18)$$

yields a singular point. The general direction that the phase trajectory traces out in time can be obtained at any point on the phase plane (i.e. the x-y plane) by noting the signs of P and Q; thus:

$$P > 0, \quad \text{and} \quad Q > 0, \quad \text{then} \nearrow,$$

$$P > 0, \quad \text{and} \quad Q < 0, \quad \text{then} \nwarrow,$$

$$P < 0, \quad \text{and} \quad Q > 0, \quad \text{then} \searrow,$$

$$P < 0, \quad \text{and} \quad Q < 0, \quad \text{then} \swarrow.$$

The characteristic curves, Equations VII.18, thus divides up the phase plane into four regions in each of which all phase trajectories have the same general direction.

It is found that the characteristic of a singular point (except for the closed curve case) is entirely determined by the slopes of the characteristic curves, Equations VII.18, at the singular point, provided the sign of Q above $Q(x,y) = 0$ is known and the sign of P to the right of $P(x,y) = 0$ is known. In particular, it will be assumed here that $\partial Q/\partial y < 0$ and $\partial P/\partial x > 0$ in the neighborhood of the singularity. A reflection in the x or y variables can always guarantee this condition. It is also assumed that

$P(x,y) = 0$ is single valued function of x and that $Q(x,y) = 0$ is a single valued function of y in the neighborhood of the singularity.

Phase trajectories in the neighborhood of a singular point can behave in any one of the following ways:

1.  pass directly into the singular point without spiraling,
2.  pass directly out from the singular point without spiraling,
3.  be deflected by the singularity,
4.  be deflected without crossing the characteristic curves,
5.  spiral into the singularity,
6.  spiral out from the singularity, or
7.  be a closed curve about the singularity.

The various possibilities are indicated in Figure 20A. Cases 1 and 5 are stable with a unique solution for x and y as $t \rightarrow \infty$, and case 7 is stable with an oscillatory solution. Ordinarily a stable solution would be expected to be of the type indicated in case 5.

### Expansion About the Singular Point

Probably the most important aspect of any phase diagram is the nature of the singular points that occur. In particular, the nature of a singularity may be ascertained by expanding $P(x,y)$ and $Q(x,y)$ about the singular point yielding to first powers in x and y

$$dy/dt = a_o + a_1x + a_2y ,$$
$$dx/dt = b_o + b_1x - b_2y , \qquad (VII.19)$$

where $a_1$ and $b_2$ are positive, since $P > 0$ to the right of $(Px,y) = 0$ and $Q < 0$ above $Q(x,y) = 0$. Shifting the origin to the singular point,

$$dy'/dt = a_1x' + a_2y' ,$$
$$dx'/dt = b_1x' - b_2y' , \qquad (VII.20)$$

where

$$x' = x + (a_ob_2 + a_2b_o)/(a_1b_2 + a_2b_1) ,$$
$$y' = y + (b_1a_o - a_1b_o)/(a_1b_2 + b_1a_2) .$$
$$(VII.21)$$

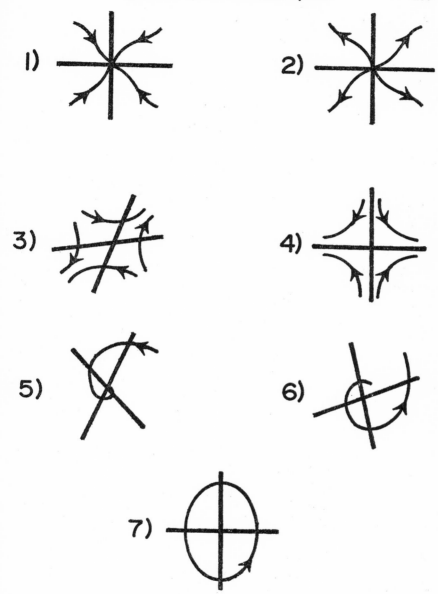

Figure 20A. Various possible phase trajectories in the neighborhood of a singular point.

Making scale changes in the variables, Equations VII.20 become

$$dm/dt' = n + am \, ,$$
$$dn/dt' = bn - m \, , \qquad (VII.22)$$

where

$$t' = \sqrt{a_1 b_2}\, t \, , \qquad m = \sqrt{a_1}\, b_2 y' \, , \qquad n = a_1 \sqrt{b_2}\, x' \, ,$$
$$a = a_2 \sqrt{a_1 b_2} \, , \qquad b = b_1 \sqrt{a_1 b_2} \, . \qquad (VII.23)$$

Assuming a time variation $e^{pt'}$ ,

$$p = (a + b)/2 \pm i[1 - (a - b)^2/4]^{\frac{1}{2}} \, . \qquad (VII.24)$$

From Equations VII.22 the slope of $P(n,m) = 0$ and $Q(n,m) = 0$ are given respectively by

$$S_p = -1/a \, , \qquad S_q = b \, . \qquad (VII.25)$$

If the sign of the quantity under the radical in Equation VII.24 becomes negative, there are two possible real values of p, indicating a point of avoidance for the the singular point. If the quantity under the radical in Equation VII.24 remains positive, it represents an oscillatory component; and whether the solution spirals in or out depends upon the sign of the first term in Equation VII.24. In particular,

$$|S_q + 1/S_p| \begin{Bmatrix} > 2 \\ \le 2 \end{Bmatrix} \quad \begin{matrix} \text{phase trajectories avoid point,} \\ \text{phase trajectories spiral in or out.} \end{matrix}$$
$$(VII.26)$$

If the latter condition is true, Equation VII.24 yields from the first term, using Equations VII.25:

$$S_q > 1/S_p \quad \text{phase trajectories spiral out,}$$
$$S_q < 1/S_p \quad \text{phase trajectories spiral in,} \qquad (VII.27)$$
$$S_q = 1/S_p \quad \text{phase trajectories are closed curves.}$$

A conversion of the slopes in the n-m plane back into the x-y plane may be achieved by multiplying $S_q$ and $S_p$ by $\sqrt{a_1/b_2}$. Geometrical constructions are facilitated by noting that

$$S_q = \tan \theta_q \, , \qquad 1/S_p = \tan \theta_p \, , \qquad (VII.28)$$

where $\theta_q$ is measured from the n axis and $\theta_p$ is measured from the m axis.

## A Theorem to Determine the Nature of a Singular Point

Considering Stokes theorem,

$$\int_A (\nabla \times \mathbf{B}) \cdot \mathbf{n} \, da = \oint_C \mathbf{B} \cdot \mathbf{ds}, \tag{VII.29}$$

where C is a closed curve about the area A, and **B** is a well-behaved vector on A, and letting $B_x = -P(x,y)$ and $B_y = Q(x,y)$, the following identity is obtained

$$\oint_C (Q \, dy - P \, dx) = \int_A (\partial P/\partial y + \partial Q/\partial x) \, dxdy. \tag{VII.30}$$

Along any phase trajectory the integrand on the left side is zero. If a phase trajectory is closed then the left side is zero and the right side must also be zero, and the integrand on the right must be zero or change sign, a theorem by Bendixson (1901).

If the phase trajectory is a spiral then a closed path may be obtained by following the phase trajectory from the curve $P(x,y) = 0$ counterclockwise around the singularity back to $P(x,y) = 0$ and then along $P(x,y) = 0$ to the starting point as indicated below: Fig. 20B. For this choice of closed path the

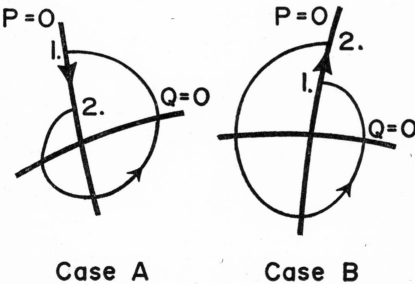

## Case A          Case B

Figure 20B.

integrand on the left of Equation VII.30 is just the contribution along $P(x,y) = 0$ between points 1 and 2; or

$$\int_1^2 Q \, dy = \int_A (\partial P/\partial y + \partial Q/\partial x) \, dxdy . \qquad (VII.31)$$

For $P > 0$ to the right of $P(x,y) = 0$ and $Q < 0$ above $Q(x,y) = 0$ the statement of the theorem is then:

*If the integral on the right of Equation VII.31 is negative, the phase trajectory spirals in and the solution is stable. If the integral on the right of Equation VII.31 is positive, the phase trajectory spirals out and the solution is unstable.*

In proof it may be noted that if the integral on the right of Equation VII.31 is negative, then (since Q is negative on the left) the direction of integration on the left is positive and the phase trajectory is spiraling inward as shown in case B above. If, however, the sign of the integral on the right of Equation VII.31 is positive, then the direction of integration on the left must be negative and the phase trajectory is spiraling outward as shown in case A above.

### Failure of the Volterra Model

Returning to the problem of interest here, Equations VII.16, VII.17 and VII.31 yield

$$\int_1^2 Q \, dy = \int_A (cx - e + a - by) \, dxdy . \qquad (VII.32)$$

From the first of Equations VII.16 it may be seen that $Q < 0$ above $Q(x,y) = 0$. Setting Equations VII.16 equal to zero, the integrand on the right of Equation VII.32 at the singular point equals $(exy_o/\underline{y} + ax_o/\underline{x})$ where $\underline{x} > 0$ and $\underline{y} > 0$ are the values of x and y at the singular point. For points sufficiently near the singular point the integrand on the right is therefore positive; and according to the theorem stated above the phase trajectory spirals outward and yields an unstable solution as indicated in Figure 21. This same conclusion may also be derived by expand-

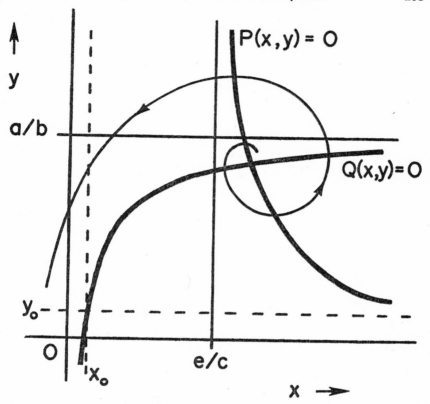

Figure 21. Predator y and prey x relation for the Volterra theory when a minimum population ($x_o$ and $y_o$) for survival are included. Extinction of the predator always occurs.

ing P and Q about the singular point and following the traditional but lengthy analysis as explained above.

There are two possible outcomes here: either the predator becomes extinct and the prey continues to live, or the predator and prey both become extinct, depending upon whether the y or x axis is first crossed by the phase trajectory. In either case the slight realistic modification of the Volterra problem by including a minimum population for reproduction always results in extinction of the predator, indicating that the original Volterra problem is not realistic. It may be concluded that actual populations cannot ordinarily be predicted by the Volterra Equations VII.12.

Experimental observations (e.g Gause, 1934; Neyman, 1956; Utida, 1957; Leigh, 1969; Dawson and King, 1971; Ayala, 1971), while frequently showing some oscillations, generally do not support the Volterra model.

In order to generate a more realistic theory which allows for the possibility of stable solutions, it is necessary to choose $P(x,y)$ and $Q(x,y)$, Equations VII.17, more realistically, even if the resulting mathematics becomes more complicated.

### Reproduction Rate

It appears that the actual birthrate, and therefore reproduction rate, of a species in nature rarely is directly coupled to the amount of food available, as required by the Volterra theory. While a species may sometimes adjust its birthrate in proportion to the food available (Lack, 1969), the ability to predict the food supply available for the offspring is not apparently the usual situation.

It appears that the greatest security against extinction is afforded by species that reproduce at a rate independent of the population. For example, a species that must nest in a hollow tree may attain a reproduction rate which depends only on the number of hollow trees and not upon the population of the species. Similarly, birds that have nesting territories of a fixed size will cover an area with the same number of breeding pairs each year and will tend to produce the same number of births each year no matter what the size of the population or the amount of food available (assuming a sufficiently large population and a sufficient food supply). Such birds have evolved complex morphological structures and behavior patterns in order to enforce a fixed size for each territory. The size of a territory is probably no larger than that just needed to feed the family. More discussion of territoriality is included in Chapter X.

If the population of birds that maintain nesting territories becomes too small, the nesting territories cannot all be filled and the number of births will then depend upon the population. In particular, for a dilute population with plenty of food the number of births should follow the Malthusian law. As observed in nature, the population should be above some fixed number (perhaps a mated pair) in order for any births to occur at all.

To take into account all of these features mentioned above (while neglecting others) the birthrate $(dy/dt)_r$ may be approximated by the following function of the population y:

$$(dy/dt)_r = c(y - y_0)/(y + y_1) , \qquad (VII.33)$$

where c, $y_0$ and $y_1$ are constants. It is assumed that $y_0 < < y_1$.

The birthrates goes to zero if y falls to the small value $y_0$. For y greater than $y_0$, but still much smaller than $y_1$, the birthrate is given approximately by

$$(dy/dt)_r = c'(y - y_0) , \qquad (VII.34)$$

where $c' = c/y_1$, which would yield the Malthusian population growth if no other factors were involved. For y somewhat larger, but still much smaller than $y_1$, second order terms in y may be included to give

$$(dy/dt)_r = c'(y - y_0)(1 - y/y_1) , \qquad (VII.35)$$

which, neglecting the small quantity $y_0$, reduces to the example of Verhulst, Equation VII.7, already considered. It should be noted, however, that the *density dependent* term is not assumed to arise necessarily from external effects, but, instead, it reflects a more realistic choice of the birthrate which includes possible self-regulatory mechanisms.

Finally, for large values of y the reproduction rate becomes a constant independent of the population; or Equation VII.33 yields

$$(dy/dt)_r = c . \qquad (VII.36)$$

It will be indicated below that a species following this reproduction rate tends to have a greater survival potential; consequently, evolution would tend to pick species which follow this reproduction rate, such as a territorial species.

None of the previous attempts to use realistic birthrates have allowed for this most important case of a birthrate independent of the population. Gompertz (1825) used cy ln(K/y) where c and K are constants. Rosenzweig (1971) investigated $cy(y^{-a} - K)$ where c, a and K are constants. Other formulas have been tried from time to time (eg. Gause and Witt, 1935; Smith, 1963; Watt, 1960a). May (1971) assumed a birthrate dependent upon the number of prey, $cy(1 - e^{-ax})$, where c and a are constants. No such direct coupling with the food supply is assumed here.

While a more complete analysis of reproduction rates would be desirable, the present work is limited to the mathematical model given by Equation VII.33 which, as has been shown, can be made to fit many individual circumstances (including the classical ones of Malthus and Verhulst) with the appropriate choices of the three constants c, $y_0$ and $y_1$. To indicate that further work is needed, it may be noted that a species consisting of two sexes has a reproduction rate dependent upon the rate that males and females can find each other in order to mate. The highly evolved auditory and olfactory sensitiveness of certain insects to achieve mating indicates the importance of the problem.

Before ending this section it should be noted that one of the reasons that the Volterra reproduction rate proportional to the predator population times the prey population, xy, appears reasonable is that the rate of combination of two compounds to form a third is ordinarily set equal to a constant times the product of the concentrations of the two compounds. As a consequence, it is reasoned that the material necessary to produce the offspring of a species comes from a product of the food supply and the population of the species. This view is in error, however, not only because it fails to fit the observed facts concerning animal populations, but also because it fails to fit the facts concerning the rates of biochemical combinations. When complicated chains of chemical events are involved, for example, in the little understood area of catalysis, the rate of formation of a compound from two compounds may depend only upon the concentration of one of the compounds; or for some concentrations of the compounds the rate may be entirely independent of the concentration of either compound. For example, if the time between collisions of the reacting molcules is much shorter than the time for the product molecule to decay into a stable state, then the rate of chemical combination becomes entirely dependent upon the rate of decay into the stable state and not upon the rate of collision. In particular, the birthrate of bioorganic molecules of even moderate complexity is an extremely difficult subject which transcends the law of mass action (e.g. Nicolis and Babloyantz, 1969; Higgins, 1967; Zaikin and Zhabotinsky, 1970).

### Death Rate

The death rate of a population clearly depends upon a very large number of factors. In order to limit the mathematical problem to something manageable, it will be assumed here that the death rate is a sum of the death rates due to various individual causes. This assumption, consequently, ignores the very real coupling between the various causes of death. For example, individuals weakened by malnutrition must have an increased death rate due to almost all other causes.

DEATH RATE DUE TO STARVATION. The death rate due to starvation has two outstanding characteristics: When the quantity of food available goes to zero (or some value too small to keep even one member of the population alive), the death rate due to starvation goes to infinity (the time for one member of the population to die being assumed to be essentially zero). And when the quantity of food gets very large, the death rate due to starvation may be assumed to go to zero. In this case the rate at which food is eaten will be proportional only to the population and will be independent of the food supply. In order to take into account these two outstanding characteristics, the death rate due starvation, $-(dy/dt)_s$, is assumed to be given by the approximation

$$-(dy/dt)_s = by/(x - x') , \qquad (VII.37)$$

where x is the amount of food (or the population of prey) and b and x' are constants. The constant x' indicates the amount of food which cannot be found or consumed by even a starving population. In some cases x' must be placed equal to zero, meaning that all available food (or prey) can be found and can be consumed. The constant x' in other cases may be quite large.

This situation occurs when the search for food is unsuccessful even when the food supply may be plentiful. Under these circumstances a population may become extinct even in the presence of a large food supply. Such an apparent shortage of food is referred to by Andrewartha (1961b) as a *relative shortage* of food.

A more complete and realistic analysis of actual death rates due to starvation must be postponed; the present analysis is limited to a consideration of the simple mathematical model presented by Equation VII.37.

DEATH DUE TO OLD AGE. As is well known, immediate causes of death never include old age. However, there is usually obvious deterioration with age and an exponentially increasing death rate due to all immediate causes with age, so that statistically old age may be considered a primary cause of death. To a first approximation it will be assumed here that the population of interest has an age distribution which does not change with time. Under these assumptions the death rate due to old age may be approximated by

$$-(dy/dt)_a = a'y, \qquad (VII.38)$$

where $a'$ is a constant.

Since the life span of an individual member of a population may be long compared with time lapses of interest (as is true for the human population), Equation VII.38 may not be adequate. The time variation of the total population may actually depend upon the age distribution of the population. This extremely interesting area of research (comparable to Volterra's hereditary problem) will not be considered here, however; only the first order approximation, Equation VII.38 will be included.

DEATH RATE DUE TO A COMMON HAZARD. If all of the individuals in a population are subject to the same hazard, such as being struck by lightning, or being bitten by a rabid animal, then the death rate due to this common hazard is proportional to the population; thus,

$$-(dy/dt)_h = a'' y, \qquad (VII.39)$$

where $a''$ is a constant.

DENSITY DEPENDENT DEATH RATE. As indicated previously a death rate varying as the density of the population, $y^2$, probably reflects neither the effect of communicable diseases nor the effect of crowding accurately. For mathematical completeness, however, such a term may be included; and, if it is not important, the coefficient may be taken as zero; thus,

$$-(dy/dt)_d = e\,y^2\,, \qquad\qquad (\text{VII.40})$$

where e is a constant.

DEATH RATE CAUSED BY A PREDATOR. A predator is regarded here as any other species which kills members of the population of interest. Besides species that eat their prey, this definition includes parasites which are lethal or contribute to the population's death rate. As mentioned before, for low densities of predator z and prey y the rate at which predator can catch its prey is proportional to a product of the two populations yz. However, when the prey population gets very large compared with the predator population, then the number of prey taken becomes essentially dependent only upon the number of predators, each predator being able to consume food at only some fixed maximum rate.

To represent these ideas mathematically the death rate due to a predator is approximated by

$$-(dy/dt)_p = f\,yz/(y + y_2)\,, \qquad\qquad (\text{VII.41})$$

where f and $y_2$ are constants. The constant f equals the maximum number of prey that one predator can eat per unit time. While the present account will be limited to Equation VII.41, it may be noted that a more complete account would have to take into consideration many variables. This death rate by predation, Equation VII.41 was originally investigated by Holling (1959 and 1965). Gauss (1934) assumed a predation rate proportional to $y^{1/2}z$; Rosenzweig (1971) considered $y^{\gamma}z$ where $\gamma$ is a constant, $0 < \gamma < 1$; and other formulas have been treated by Watt (1960b) and Goel *et al.* (1971).

NET DEATH RATE DUE TO ALL CAUSES. The net death rate due to all causes is approximated here by the sum of death rates due to starvation, Equation VII.37; the death rate due to old age, Equation VII.38; the death rate due to a common hazard, Equation VII.39; the death rate due to density dependent factors, Equation VII.40; and the death rate due to a predator, Equation VII.41. Thus,

$$-(dy/dt)' = by/(x - x') + ay + ey^2 + fyz/(y + y_2) .$$
$$(VII.42)$$

where b, x', a = a' + a'', e, f and $y_2$ are constants, x is the amount or population of food and z is the  population of predators.

### Net Time Rate of Increase in Population

The net time rate of increase in population, according to the present approximations, is then given by the reproduction rate, Equation VII.33, less the net death rate due to all causes, Equation VII.42; or

$$dy/dt = c(y - y_e)/(y + y_1) - by/(x - x') -$$
$$ay - ey^2 - fyz/(y + y_2) . \qquad (VII.43)$$

Because this expression has nine adjustable constants, c, $y_o$, $y_1$, b, x', a, e, f and $y_2$, as well as two continuous variables x (the food supply) and z (the population of predators feeding on y), it should be expected that this expression can be adjusted to fit many actual situations. The general case is too difficult to handle analytically (although numerical explorations with a computer need not be ruled out); however, there are many interesting special cases where the order of magnitude of the coefficients may be estimated.

If the population y feeds on a number n of other species with populations $x_1$, $x_2$, . . ., $x_n$ and if a number of m of other species with populations $z_1$, $z_2$, . . ., $z_m$ prey on the population y, then Equation VII.43 may be generalized to read

$$\frac{dy}{dt} = c\frac{y - y_o}{y + y_1} - y \sum_{i=1}^{n} \frac{b_i}{x_i - x'_i} - ay - ey^2 - y \sum_{i=1}^{m} \frac{f_i z_i}{y + y_{2i}} .$$
$$(VII.44)$$

In every ecosystem there are always a number N of species (or nonliving food supplies) which need to all be considered. Each of the species present will have a population which will be governed (to within the present approximations) by an equation of the form of Equation VII.44. This leads to a set of N simultaneous differential equations,

$$\frac{dy_j}{dt} = c_j \frac{y_j - y_{oj}}{y_j + y_{1j}} - y_j \sum_{\substack{i=1 \\ j \neq i}}^{N} \frac{b_{ij}}{y_i - x'_{ij}} - a_j y_j - e_j y_j^2$$

$$- y_j \sum_{\substack{i=1 \\ j \neq i}}^{N} \frac{f_{ij} y_i}{y_j + y_{2ij}} \qquad , \quad (VII.45)$$

where the constants $c_j$, $y_{oj}$, $y_{1j}$ and $e_j$ are peculiar to the $j$th population, while the constants $b_{ij}$, $x'_{ij}$, $f_{ij}$ and $y_{2ij}$ represent coupling between $j$th and $i$th populations. These Equations VII.45 have been presented here, not because they are going to be solved in the general form, but because it is important to note that the present mathematical models are applicable (at least in principle) to the entire ecosystem and computers may be able to eventually explore such a model for an ecosystem numerically. Lotka (1924c) presented a similar set of equations to indicate the same idea.

In even more general terms, an ecosystem of N interacting species, $y_1, y_2, \ldots, y_N$ may be described by N diffierential equations of the form

$$dy_j/dt = P_j(y_1, y_2, \ldots, y_N) . \qquad (VII.46)$$

Empirically the central problem is to discover the appropriate functions $P_j$. As has been indicated, Lotka and Volterra did not choose very realistic expressions for their P's. The particular choice made here, Equation VII.45, while undoubtedly more realistic, probably leaves much to be desired.

The general problem will now be abandoned in favor of considering some more tractable limiting cases. These limiting cases, however, will yield some insight into the behavior of populations.

### Food Supplied at a Constant Rate

In this section the problem of a population y being provided with a single food x at a constant rate will be considered. The amount of food will be assumed to be reduced at a rate propor-

tional to the rate of eating by the population of interest and also
at a rate proportional to the amount of food. The latter allows
for the possibility of passive competition, since other species may
also be eating the food in order that the rate that the food dis-
appears is proportional to $a_o x$, where $a_o$ is a constant. It will also
be assumed for the present special case that the population y
has no predators. In this case the problem reduces to a two-
population problem. In particular, Equations VII.45 reduce to
the pair of differential equations

$$dy/dt = c_1(y - y_o)/(y + y_1) - by/(x - x') - a_1 y - ey^2 ,$$
$$dx/dt = c_o - a_o x - fxy/(x + x_2) , \qquad (VII.47)$$

where $c_1$, $y_o$, $y_1$, b, $x'$, $a_1$, e, $c_o$, $a_o$, f and $x_2$ are constants.

The present case of a population which has no effect upon
the rate that food is supplied occurs commonly in nature. For
example, detritus feeders must wait for dead plant products to
fall; they do not influence in any way the rate detritus is deposited. Similarly carrion eaters and coprophagous organisms
are supplied with food at a rate independent of their populations. The amount of sunlight received per unit area on the
earth's surface remains fixed, so that the totality of green plants
are supplied with their essential food, i.e. sunlight, at a constant
rate independent of the total population of green plants.

In addition there are many organisms that have only a very
slight effect upon the rate that food is supplied to them, and
which, therefore, to a good first approximation fit the present
special case. Animals that live off of the expendible parts of
plants fit the present category. The codlin moth eats apples
which in no way affects the number of apples produced by the
tree the following year. The codlin moth does, however, interfere slightly with the dispersal of seeds and thereby does interfere slightly with the reproduction rate of apple trees. Bees,
feeding on the nectar and pollen of flowers, do not threaten the
life of the plant, and they apparently have a positive effect on
the reproduction of the plants. Most parasites, it would seem,
do not harm their hosts appreciably; consequently, most parasites apparently have little effect upon the rate that food is supplied to them.

The present case would also fit slow-moving predators that

have difficulty in catching fast-moving prey. Such predators will only catch and eat the old, lame and sick prey and will have no appreciable effect on the healthy prey population which is reproducing. Such slow predators will also have difficulty in finding prey to capture and will, consequently, kill only a few with only a very small effect on the rate that food is supplied to them. Similarly, the present section fits those predators that do not kill their prey, but instead merely extract a small amount of tissue from their prey. A mosquito seldom kills a man or, in fact, has any appreciable effect upon the total population of large animals upon which it preys.

It has already been mentioned that species responsible for large energy and mass transfer in the biosphere are apparently, from theoretical considerations, those that do not affect their food supply. In particular, an ecosystem which evolves toward a time average maximum biomass is an ecosystem which guarantees a constant food supply for such a biomass (*see* Chapter VIII).

Naturally, all species do not fit the present case, and there are spectacular deviations in nature. For example, the moth, *Cactoblastis cactorum*, which feeds on the prickly pear, *Opuntia* (Dodd, 1936) in Australia destroys the cactus completely within the area in which the moth becomes established. This example might indicate a smaller standing biomass than would be maintained without the moth; but, since both species are not native to Australia, it probably represents more of a *test tube* case than a natural example.

A SPECIES WITH A CONSTANT BIRTH RATE AND NO DENSITY DEPENDENT DEATH RATE. For the present special case Equations VII.47 reduce to the simpler set of equations

$$dy/dt = c_1 - by/(x - x') - a_1y,$$
$$dx/dt = c_o - a_ox - fxy/(x + x_2).$$ (VII.48)

Under ordinary circumstances in nature the density dependent death rate is probably not important. However, such a term should be retained perhaps to explain laboratory situations and to account for some unusual occurrences such as gives rise to the forced migration of lemmings, rats and squirrels.

To analyze the nature of the solutions to Equations VII.48

$P(x,y) = 0$, and $Q(x,y) = 0$, Equations VII.17, may be plotted on a phase diagram, where

$$P = -a_1 [(y - c_1/a_1)(x - x' + b/a_1) + bc_1/a_1]/(x - x') = 0 ,$$
$$Q = -f [xy - (c_o - a_ox)(x + x_2)/f]/(x + x_2) = 0 ,$$
$$(\text{VII.49})$$

where $P(x,y) = 0$ is seen to be an equilateral hyperola. Provided $c_1$ and $c_o$ are always large enough, or comparable to the other terms on the right of Equations VII.48, the solution is very stable; both the food population and the population of interest

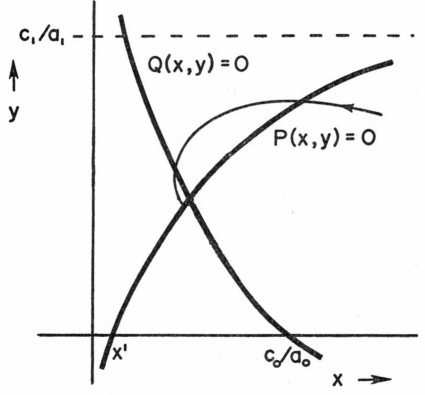

Figure 22. A population y supplied at a constant rate with food x and with a constant birth rate, Equations VII.48. This case leads to very rapid spiraling into the point of stability.

spiral rapidly into the singular point as indicated in Figure 22. Extinction can occur only in the unlikely event that x, due to extraneous reasons, becomes less than x', as indicated by the region in Figure 22 left of x'.

Because of the extreme stability of the present case, where very little oscillation can occur after a displacement from equilibrium (i.e. the singular point), the present case represents an ideal toward which a population might evolve in order to insure survival. Thus, evolution might tend to favor a species which exercises population control by maintaining a fixed reproduction rate (if such self-control is necessary) and which does not directly affect the supply of food upon which it depends.

The present discussion is limited to a two-population problem. For three or more populations there may be other stabilizing mechanisms which come into play. In a complex ecosystem a predator may have a direct effect on its food supply which could have the stabilizing effect of keeping the prey population in check, so that the survival potential of the prey population is increased.

A SPECIES WITH A GENERAL REPRODUCTION RATE AND FOOD SUPPLIED AT A CONSTANT RATE. It is of interest to examine the general case for reproduction where $y_0$ and $y_1$, Equations VII.47, are not taken to be zero. In particular, the equations of interest in the present instance are

$$dy/dt = c_1(y - y_0)/(y + y_1) - by/(x - x') - a_1y,$$
$$dx/dt = c_0 - a_0x - fxy/(x + x_2),  \qquad (VII.50)$$

where again the density dependent death rate has been ignored here. The importance of the present case may be indicated by recalling that the Volterra problem was revealed as unstable when a realistically small finite value of $y_0$ was included. Also the theory in the previous section was so stable that under no circumstances, except the very unlikely event of x becoming less than x', could the population become extinct; and yet populations do become extinct under apparently some even relatively mild circumstances.

To examine the behavior of the solution for the present case,

Equations VII.50, the first quadrant of the phase diagram may again be considered where Q is identical to the second of Equations VII.49 while P may be written in the form

$$P = c_1 y \left[ \frac{y - y_o}{y(y + y_1)} - \frac{a_1}{c_1} - \frac{b/c_1}{x - x'} \right] = 0 . \quad (VII.51)$$

These expressions, the second of Equations VII.49 and Equation VII.51 are indicated in Figure 23. Horizontal asymptotes for $P(x,y) = 0$ occur approximately at $y = c_1/a_1$ and $y = y_o$ (assuming $a_1 y_1/c_1 < < 1$ and $a_1 y_o/c_1 < < 1$). $P(x,y) = 0$ in the first quadrant has a minimum value of x at approximately $x = x' + b y_1/c_1$ and $y = \sqrt{y_o y_1}$ . There are two singular points in the region of interest. The upper singularity is a stable point, while the lower point is a point of avoidance. There is a large region in the phase diagram where the phase trajectory leads to stability by spiraling into the upper singular point. A portion of the phase diagram in the first quadrant, shaded in Figure 23, leads to the extinction of the population. Curves with the directions indicated show the behavior of a phase trajectory occurring in the region where the curve is drawn.

If a species by some extraneous circumstances is placed below $y = y_o$ it becomes extinct. The species also becomes extinct if it finds itself to the left and below some critical path (indicated by C in Figure 23 which passes near the point $x = x' + b y_1/c_1$, $y = \sqrt{y_o y_1}$ . In this latter region the rate of reproduction is too small to make up for the losses due to starvation and the species becomes extinct. To the right and above this critical path for $y < y_1$ the population is sufficiently large to be able to reproduce itself and extinction does not occur.

It should be noted that the actual time rate that a phase trajectory may be traced out is not being considered here. Estimating actual time variations (except possibly very near the point of equilibrium) is beyond the present analysis.

The feeling and judgment of many observational ecologists is that many more individuals of a species than the proverbial mated pair are ordinarily necessary to insure survival. It has been estimated, for example, that the promiscuous killing of whales to the point where it will no longer be of immediate economic ad-

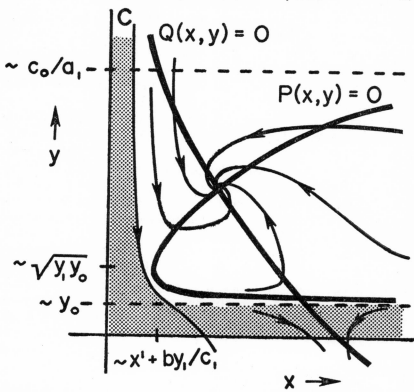

Figure 23. Case of a population y with imperfect population control supplied with food at a constant rate, Equations VII.50. The shaded area leads to extinction and the unshaded area to stability.

vantage to the last surviving individual whaler will leave a population of only a few thousands of the most threatened species, which will be too small for survival. The present analysis indicates that even a seemingly large population, if it falls to the left and below the critical path C, sketched in Figure 23, will become extinct. The present theory, thus, bears out mathematically the usual expectations. It might be thought that the present theory does not actually predict extinction because it concerns an unlikely region in the phase diagram where the population becomes very large or the amount of food very small; but the amount of food per individual of the population need not be small. In particular, it may be noted that any point on a straight line joining the origin to the point of stable equilibrium provides as much food per individual as at the point of stable equilibrium.

Similarly, it might be thought that a serious reduction in the population of a species would permit a marked increase in the food supply available and, consequently, a subsequent increase in the population. But this is hardly ever the actual case, since there are usually a large number of species competing for the same food supply. Ordinarily it may be expected that if a particular population decreases, then other species, eating the same food, will rapidly take care of any excess food that might develop. The amount of food available per individual of the species of interest, therefore, tends to remain about the same independent of the size of the population.

INCLUDING A DENSITY DEPENDENT DEATH RATE. It is now possible to consider the general case for this section by letting e in Equations VII.47 be nonzero. It may be seen that the additional term does not appreciably affect the function $P(x,y) = 0$ as given by Equation VII.51 in the region of interest. Consequently, Figure 23 may also be viewed as representing the general case of a population feeding on a supply of food which is supplied at a constant rate.

## The General Two-population Problem

In this section the special requirement that food be supplied at a constant rate is no longer considered. As already indicated, the inclusion of a minimal population below which a population cannot reproduce led to a necessary extinction of the predator population in the Volterra problem. Thus, it is clear that a more general consideration of the prey population, or food supply, is necessary. In particular, the present section will investigate a food supply with a reproduction rate given by Equation VII.33, x replacing y, in order to see whether or not it provides for the possibility of the survival of the predator y.

In order to reduce the present approximate analysis to a two-population problem, it will be assumed that the food supply for the prey is always adequate, so that the death rate for the prey population due to starvation may be ignored (which is equivalent to letting $b = 0$ in Equation VII.43. In addition, it is necessary to assume that the predator population, the population of interest, is not preyed upon in turn. From Equation VII.43 the general two-population problem of interest here becomes

$$dy/dt = c_1(y - y_o)/(y + y_1) - by/(x - x') - a_1y - e_1y^2 ,$$
$$dx/dt = c_o(x - x_o)/(x + x_1) - a_ox - e_ox^2 - fxy/(x + x_2).$$

$$(VII.52)$$

The behavior of $P(x,y) = 0$ is essentially that displayed in Figure 23, while the behavior of $Q(x,y) = 0$ is now quite different. The two curves of interest, $P(x,y) = 0$ and $Q(x,y) = 0$, are given respectively by the dimensionless equations:

$$\underline{x} - \underline{x}' = b' [(\underline{y} - \underline{y}_o)/\underline{y}(\underline{y} + 1) - a'_1 - e'_1 \underline{y}]^{-1} ,$$
$$f' \underline{y} = (\underline{x} + \underline{x}_2)[(\underline{x} - \underline{x}_o)/\underline{x}(\underline{x} + 1) - a'_o - e'_o \underline{x}],$$

$$(VII.53)$$

where the new variables are

$$\underline{x} = x/x_1 , \qquad \underline{y} = y/y_1 , \qquad (VII.54)$$

and the constants are given by

$$\underline{x}' = x'/x_1 , \ \underline{x}_2 = x_2/x_1 , \ \underline{x}_o = x_o/x_1 , \ \underline{y}_o = y_o/y_1 ,$$
$$a'_o = a_o x_1/c_o , \ e'_o = e_ox^2_1/c_o , \ a'_1 = a_1y_1/c_1 ,$$
$$e'_1 = e_1y^2_1/c_1 , \ b' = by_1/c_1x_1 , \ f' = fy_1/c_o .$$

$$(VII.55)$$

While these curves, Equations VII.53, are simple enough in the first quadrant (the only region of interest here) to permit some general analysis, it is more illuminating to consider a partial numerical example. The numbers chosen may not be appropriate to an actual experimental or observational situation, but they serve to display the functional relationships indicated by Equations VII.53. In particular, the parameters characteristic of each species independent of the other which occur inside the square brackets in Equations VII.53 are taken to be

$$\underline{y}_o = y_o/y_1 = 10^{-2} , \ a'_1 = a_1y_1/c_1 = 0.5 , \ e'_1 = e_1y^2_1/c_1 = 0.1 ,$$
$$\underline{x}_o = x_o/x_1 = 10^{-2} , \ a'_o = a_ox_1/c_o = 0.5 , \ e'_o = e_ox^2_1/c_o = 0.1 .$$

$$(VII.56)$$

The function appearing on the right side of the first of Equations

VII.53 (which is the reciprocal of the function in the square bracket in the second of Equations VII.53),

$$F = 10\,\underline{y}(\underline{y} + 1)/(-\underline{y}^3 - 6\,\underline{y}^2 + 5\,\underline{y} - 1/10)\,, \tag{VII.57}$$

has the values given by:

| $\underline{y}$ | $F\,(\underline{y})$ | $\underline{y}$ | $F\,(\underline{y})$ |
|---|---|---|---|
| $\approx 0.022$ | $\infty$ | 0.200 | 3.68 |
| 0.025 | 11.4 | 0.300 | 4.12 |
| 0.050 | 3.75 | 0.400 | 6.39 |
| 0.100 | 3.24 | 0.500 | 9.68 |
| 0.110 | minimum | 0.600 | 18.3 |
| 0.150 | 3.37 | 0.700 | 102 |
| | | $\approx 0.705$ | $\infty$ |

Including a choice for the value of $x'_2 = x_2/x_1 = 0.5$, the values of the function $f'\underline{y}(\underline{x})$ for $Q(x,y) = 0$ become $f'\underline{y}(\underline{x}) = (\underline{x} + 0.5)/F\,(\underline{x})$, or

| $\underline{x}$ | $f'\underline{y}(\underline{x})$ | $\underline{x}$ | $f'\underline{y}(\underline{x})$ |
|---|---|---|---|
| $\approx 0.022$ | 0 | 0.250 | 0.182 |
| 0.025 | 0.0461 | 0.300 | 0.171 |
| 0.050 | 0.146 | 0.400 | 0.141 |
| 0.100 | 0.185 | 0.500 | 0.103 |
| 0.150 | 0.193 | 0.600 | 0.0600 |
| 0.200 | 0.190 | 0.700 | 0.0118 |
| | | $\approx 0.705$ | 0 |

If the two curves $P(x,y) = 0$ and $Q(x,y) = 0$ fail to intersect, the population y must become extinct, there being no point of stable equilibrium. If the two curves intersect such that the maximum of $Q(x,y) = 0$ is much lower than the upper asymptote of $P(x,y) = 0$, as indicated in Figure 24, there is still no point of stable equilibrium, and the population y becomes extinct. If the two curves intersect where the maximum of $Q(x,y) = 0$ is somewhat lower than the upper asymptote of $P(x,y) = 0$, as indicated in Figure 25, a small region in the phase diagram

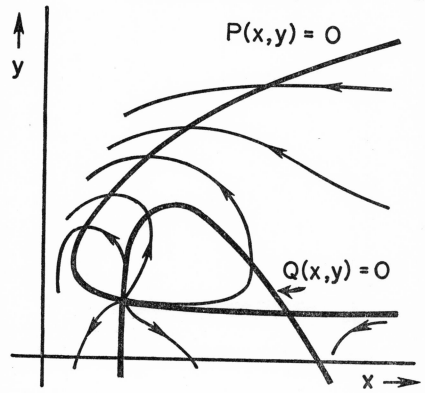

$P(x,y) = O$

$Q(x,y) = O$

y

x →

Figure 24. The general case, Equations VII.52, with two singular points where survival is impossible in any region of the phase diagram due to the maximum of $Q(x,y) = 0$ being too low compared with the upper asymptote of $P(x,y) = 0$.

exists for which the population y may survive. The curved C in Figure 25 indicates the separation between the region in which survival occurs (unshaded) and the region in which extinction occurs (shaded). It should be noted that a sufficient increase in the food supply will produce a trajectory starting directly to the right of the point of stable equilibrium in a region which leads to extinction of the population y. In this case, the population, under the stimulus of a temporary increase in the food supply, increases sufficiently to eventually destroy its own food supply, resulting in its own extinction.

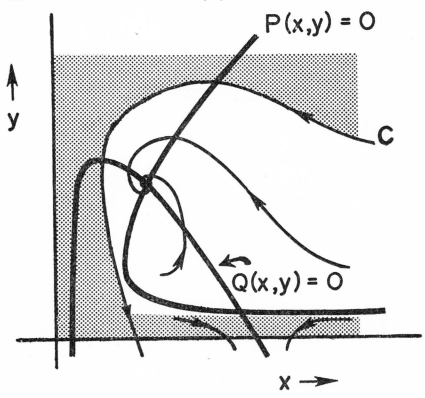

Figure 25. The general case, Equations VII.52, with two singular points where only a small region (unshaded) allows for survival due to the maximum of $Q(x,y) = 0$ being less than the upper asymptote of $P(x,y) = 0$.

When the maximum of $Q(x,y) = 0$ is sufficiently higher than the upper asymptote of $P(x,y) = 0$, then a large region of stable equilibrium results, as indicated by the unshaded portion in Figure 26. This case is analogous to that shown in Figure 23. If the two curves, $P(x,y) = 0$ and $Q(x,y) = 0$, intersect at four singular points, as indicated in Figure 27, there is no appreciable change in the situation, since the two additional points yield one point of avoidance and the other of outward spiraling near the $x = 0$ axis. The fact that the slope of the phase trajectories become infinite along the line $x = x'$ has not been indicated in these Figures 24, 25, 26 and 27; this fact should be kept in mind for some special cases.

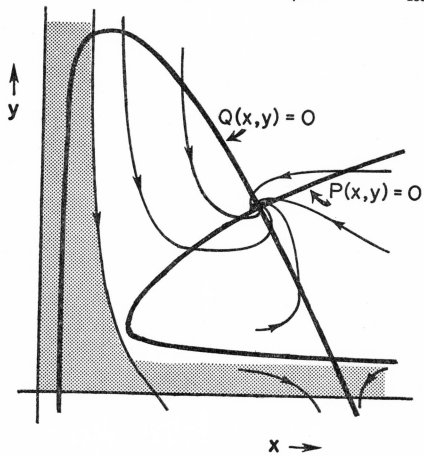

Figure 26. The general case, Equations VII.52, with two singular points where a large region (unshaded) allows for survival due to the maximum of $Q(x,y) = 0$ being large compared with the upper asymptote of $P(x,y) = 0$.

By inspection of Figures 24, 25, 26 and 27 it becomes apparent that conditions favorable for the survival of a population y (under the present assumptions) tend to be for $Q(x,y) = 0$ as high as possible and $P(x,y) = 0$ as low as possible and displaced toward the left. It is now of interest to consider the values of the parameters which permit such favorable conditions for survival.

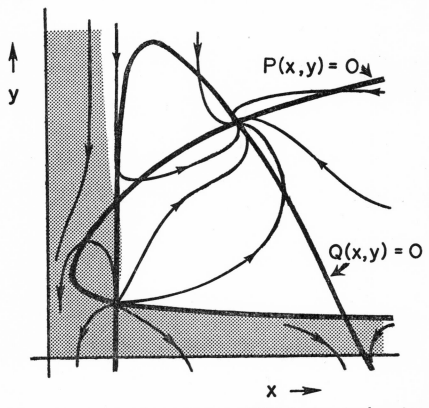

Figure 27. The general case, Equations VII.52, for four singular points, one stable, two points of avoidance, and one yielding outward spiraling. There is little change from the situations indicated in Figures 25 and 26.

The parameters that indicate coupling between the two populations y and x are $x_2$, $x'$, b and f, Equations VII.52. An increase in the value of $x_2$ tends to raise the curve $Q(x,y) = 0$. This indicates that a population y will tend to adapt evolutionarily to increase $x_2$ or to eat under a condition of scarcity of food (y being negligible as compared with $y_2$ in Equation VII.41).

A change in the value of $x'$ indicates a translation of $P(x,y) = 0$ in the x direction without any change in shape, without any change in the asymptotes, and without any change in the position of the minimum as a function of y. A shift toward the left of $P(x,y) = 0$, providing a greater survival potential, indicates that a species will tend to evolutionarily adapt to survive on a smaller supply of food, tending to utilize all sources of available food.

A change in the value of b indicates a change in the x scale for $P(x,y) = 0$, as indicated in the upper portion of Figure 28, without any change in the asymptotes or position of the mini-

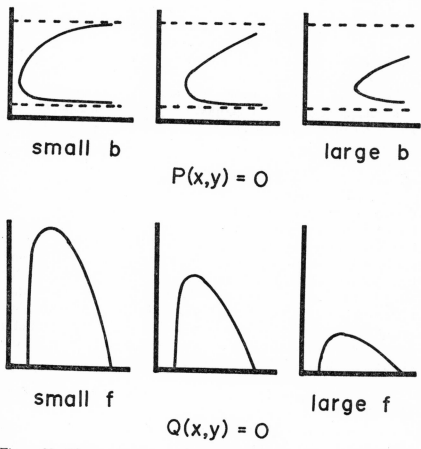

Figure 28. The upper curves indicate $P(x,y) = 0$ (the first of Equations VII.52 set equal to zero) for various sizes of b. The lower curves indicate $Q(x,y) = 0$ (the second of Equations VII.52 set equal to zero) for various sizes of f. The ordinates indicate y and the abscissas indicate x.

mum as a function of y. The increased survival potential for small values of b indicates that a species should evolve toward a decrease in the death rate due to starvation, Equation VII.37. Presumably this can occur only for an increase in the rate that food energy is gathered in comparison to the rate that energy must be expended to gather food, i.e. an increase in the food gathering efficiency.

A change in the value of f indicates a change in the y scale for $Q(x,y) = 0$, as indicated in the lower portion of Figure 28, without any change in the x position of the zero values or the maximum value. An increase in the maximum of $Q(x,y) = 0$, which leads to the greater survival potential, requires a decrease in f. This means, from Equation VII.41, that a population y should evolve toward a minimum amount of food consumed, or a maximum utilization of food consumed. Since population here is measured in terms of carbon mass, this means that evolutionary selection should favor the more ecologically efficient organisms (a conclusion derived from general theoretical considerations in the next Chapter).

Apart from the parameters that couple the two populations, the survival potential of the population y depends markedly upon the magnitude of the upper asymptote of $P(x,y) = 0$. If there is no restraint on the reproduction rate of the population y, then y may be neglected compared with $y_1$ in Equation VII.33 and in Equation VII.52. This leads to a much larger asymptote for $P(x,y) = 0$, which changes the situation from essentially that indicated in Figure 26 to that indicated in Figure 25. Considering $Q(x,y) = 0$ as a fixed curve, it is clear that the survival of a population y is increased by a lowering of the upper asymptote of $P(x,y) = 0$.

## Possible Comparison of the Present Theory with Observations

Since analytic results for the present theory cannot be readily obtained, it must be assumed that computer calculations are necessary. There has been no attempt here to compare the present theory numerically with observations. However, the present theory appears to be well suited to fit actual experimental or field data because of the realistic functional relationships chosen and the large number of adjustable parameters. The present theory should be capable of providing a practical framework into which numerical observations can be fed.

## Limit Cycles

Other possible solutions to the pair of coupled Equations VII.17 involve limit cycles. This can occur, for example, when a phase trajectory that is originally outside of a particular closed curve (called the *limit cycle*) spirals into the closed curve without crossing it, while a phase trajectory originally inside the closed curve spirals out to the closed curve asymptotically as indicated in Figure 29. Thus, a unique, stable, undamped oscillating solution is finally established no matter what the initial conditions. Similarly, unstable limit cycles may also occur (Minorsky, 1962). While mathematically interesting and while of possible value in some special situations as indicated by May (1971) and Gilpin (1971) the matter is too involved to pursue any further here.

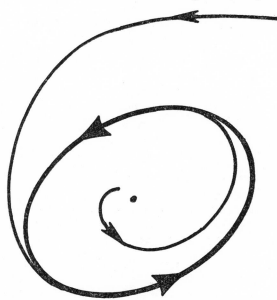

Figure 29. A diagram indicating a stable limit cycle about a point of unstable equilibrium. All phase trajectories spiral in or out to the limit cycle.

# CHAPTER VIII

# EVOLUTION IN THE PAST

THE EVOLUTIONARY PROCESS is a process that takes place in time, so that a proper delineation of various types of processes in time should be indicated here, all of them being important in one way or another in ecophysics. In particular, the following four classifications may be made:

1. *Steady state processes* which do not require an explicit specification of the time for their description. A river, for example, may be characterized by a depth, a width and a rate of flow, the time never being explicitly considered. Similarly, average metabolic rates yield steady rates of consumption of calories and oxygen, the time being unspecified.

2. *Periodic changes* which require only a specification of a period. The solar radiation impinging upon the earth has a twenty-four hour period. Practically all life on the earth shows signs of a consequent circadian rhythm. There are similarly important yearly changes of temperature and rainfall that require similar important annual adaptations of life, especially in polar regions where yearly fluctuations are the greatest.

3. *Random events* which require some designation of a probability of occurrence per unit time. The chances of rain during a day is some percent in Kansas City averaged over the year. The probability of a man dying in his forty-ninth year is a statistic of interest to insurance companies.

4. *Evolutionary processes* which continue year after year in

the same direction. Every year for thousands of years a large quantity of silt has been deposited in the Gulf of Mexico by the Mississippi River. In astronomy evolutionary changes are referred to as *secular* changes. Evolutionary processes are assumed to take place over suitably long time periods. A geologist, being used to considering long periods of time, would be inclined to consider such times as *geologic* times.

If periodic processes are averaged over many days or years, they need not be regarded as varying with a particular period such as twenty-four hours or 365 days; they may be regarded as steady state processes. Over a sufficiently long time the flux of solar energy on the earth may be taken as a steady state parameter. Random events also yield a steady state description if the average is taken over a time large compared with the mean time between occurrences. On the average Kansas City has so many inches of rain per day, a fact that may be viewed as a steady state situation. Evolutionary processes are those that still require the explicit specification of the time even after all of the long time averages have been taken into account.

### Evolution of Stars and Entropy Reduction

A star is presumably born by condensing out of clouds of interstellar gas and dust. It may be reasoned that even a uniform distribution of gas would eventually yield by random fluctuations momentary concentrations of molecules. These centers of concentration would then gravitationally attract other molecules which would be accelerated toward them. The accelerated molecules would thereby convert their gravitational potential energy into kinetic energy. Some of these accelerated molecules would collide with each other and would lose part of their kinetic energy by radiating electromagnetic radiation. Such molecules with reduced kinetic energies would then no longer be able to climb back out of the gravitational potential well into which they had fallen. They would thus, become captured. Such a capture would also increase the concentration of mass already there and would thereby increase the gravitational strength of the concentration. This process may be envisioned as continuing until mass concentrations and temperatures are attained for hydrogen to be burned by nuclear fusion. When this happens a visible star is born.

Such a condensation of gas and dust represents a decrease in the entropy of the system. For example, a galactic cloud of atomic hydrogen may be considered with a typical density of $\rho = 10^{-22}$ gm/cm$^3$. The entropy of such a tenuous gas is given by the ideal gas formula

$$S = C_v \ln T + R \ln V + S_o , \qquad (VIII.1)$$

where S is the entropy per mole, $C_v$ is the heat capacity per mole at constant volume (equal to 3R/2 for a monatomic gas), R = 2.0 cal/mole is the ideal gas constant, T is the absolute temperature, V is the volume, and $S_o$ is an arbitrary reference constant. After the gas is condensed into a star the entropy change per unit mass is given by

$$\triangle S = (3R/2M_o) \ln(T_f /T_i) + (R/M_o) \ln(\rho_i /\rho_f) , \qquad (VIII.2)$$

where $M_o$ is the molecular weight of atomic hydrogen, $V_f/V_i = \rho_i/\rho_f$, and the subscript i refers to the initial state in the galactic gas cloud and the subscript f refers to the final state in the star. Assuming a stellar density of $\rho_f = 1.5$ gm/cm$^3$ and temperature of $T_f = 10^7$ °K and assuming the temperature of the galactic cloud to be $T_i = 15°$ K, the entropy change per unit mass is $\triangle S = -62$ cal/gm ° K .

According to the second law of thermodynamics this reduction in entropy could only occur if there were an equivalent or larger increase in the entropy elsewhere in the universe. The excess entropy is carried away by the electromagnetic radiation as the star is being formed. To verify this conclusion it may be noted that almost all of the original gravitational energy of the gas cloud has to be dissipated at some sort of mean temperature $\overline{T}$. The corresponding increase in the entropy of the universe per unit mass of material condensed is then given by

$$\triangle S_u = \triangle U/M \overline{T} , \qquad (VIII.3)$$

where $\triangle U$ is the change in the gravitational potential energy for a typical star of about the mass and radius of the sun and M is the mass of the star. Assuming the original galactic cloud is spherical and that it condenses into a spherical star of radius a, the gravitational energy released is given by Equation V.22, or $\triangle U = 3GM^2/5a$. Assuming a mean stellar mass of $M = 2 \times 10^{33}$ gm, a final radius $a = 6 \times 10^{10}$ cm, and assuming a mean effective radiating temperature during the formation of $\overline{T} = 1000°$ K, the increase in the entropy of the universe becomes $\triangle S_u = 3 \times 10^4$ cal/gm ° K, which is two orders of magnitude greater than the estimated entropy decrease, indicating that condensations do not violate the second law of thermodynamics and that they are to be expected.

When such a condensation becomes dense enough and large enough to burn nuclear fuel, it reaches a period of great stability. A star on the main sequence can survive for about $10^{10}$ years without any essential changes. Such a stable star still continues, however, to decrease its internal entropy by radiating away excess entropy. Eventually, according to usual theories, it evolves into a white dwarf star where the internal matter becomes degenerate, nucleons (i.e. neutrons and protons) and electrons being crushed together at extremely high pressures. The particles arrange themselves into an orderly crystalline array of very low entropy. The whole history of stellar evolution from a galactic gas cloud to a final white dwarf is, thus, one of continuous entropy reduction. Planets, being an integral part of a particular stellar system, must also undergo a progressive entropy reduction. The living process is one of these entropy-reducing processes. Life, far from being an unusual phenomenon, merely partakes of the general entropy reduction associated with the evolution of all stars, stellar systems and the entire observable universe (neglecting the deep space radiation of about 3° K).

## The Source of Life, a Cosmological Question

The discussion in the previous section is valid provided one essential feature remains true: Excess entropy can be radiated away as electromagnetic radiation. In order for stars to be born, the excess kinetic energy of molecules and dust must be radiated away as electromagnetic radiation to provide for the gravitational capture of the particles. But if the universe were in radiative equilibrium it would be impossible for molecules and dust particles to radiate away their excess kinetic energy. Using the inverse square law for radiation and assuming a uniform density of protostars, as much electromagnetic radiation of the right sort would be received and absorbed as would be radiated and no net loss of kinetic energy could occur. This argument concerning protostars is just a statement of Olbers' paradox. Over a century ago Olbers pointed out that the night sky should be as bright as day, for, if stars are uniformly distributed in space and if radiation falls off as the inverse square law, then the amount of light received from the stars in a spherical shell of unit thickness with the observer at the center should be the same no matter what the radius of the shell, the inverse square law loss in intensity being just balanced by the number of stars in a spherical shell increasing as the square of the radius of the shell. Adding this same effect from all such spherical shells out to infinity yields an infinite amount of radiation to the observer, and night should be as bright as day. Since observations yield no such infinite flux, and night is not as bright as day, there is a paradox (Motz and Duveen, 1966a).

While a number of theories have been proposed to resolve the paradox, there is still no completely satisfactory resolution. Perhaps the cosmological red shift can account for it, since it is observed that photons lose a set fraction of their energy for each unit of distance traveled, which then implies that energy is propagated at a rate somewhat less than the inverse square law. But where does the energy go that these photons lose? Or is the red shift a Doppler red shift due to an expanding universe, as is generally assumed, where most distant objects are moving faster with respect to the observer? But whatever is the actual explanation, the important fact is that our universe does, in fact, permit systems to export a net amount of high entropy electromagnetic radiation, thereby yielding systems which continue to decrease their entropy with time. The birth of stars and the existence of life both depend upon the remarkable fact that the universe is not in radiative equilibrium and that excess entropy may always be radiated away. The fundamental source of life may be said to be this peculiar cosmological feature of the actual universe.

Viewing high-entropy photons as waste or pollution, the universe appears to accept limitless amounts of such garbage, such garbage being removed with the speed of light.

## Is Life Dependent Upon an Expanding Universe?

If it is assumed that the observed cosmological red shift (Motz and Duveen, 1966b) represents a Doppler shift and that the universe is, therefore, actually expanding with time, then the garbage radiation, which must be radiated to permit ordering processes to occur, may be regarded as being deposited in an ever-expanding plastic garbage bag. This idea may be checked by noting that, if the rate of radiation of this unwanted entropy is just matched by the rate of expansion of the universe, then numerical estimates of the temperature of deep space can be made. The energy of the radiation in deep space per unit volume is given by

$$u = a T^4 , \qquad (VIII.4)$$

where $a = 7.5641 \times 10^{-15}$ erg/cm$^3$($^\circ$K)$^4$ is a universal constant. The rate that energy must be added to deep space to maintain this energy density is then given by

$$du/dt = 3aT^4r^{-1}dr/dt = 3aT^4H , \qquad (VIII.5)$$

where the fractional increase in size $r^{-1}dr/dt$ with time has been set equal to the Hubble constant H. In addition, light in the universe continuously loses energy by the cosmological red shift. The fractional rate of loss of this energy per unit time per unit volume just equals the Hubble constant. Thus, in order to maintain steady state equilibrium this additional loss requires replacing the factor 3 appearing in Equation VIII.5 by a 4. This rate must then be matched by the radiation of energy per unit time from galaxies per unit volume s. The estimated temperature of deep space is then given by

$$T = (s/4aH)^{1/4} . \qquad (VIII.6)$$

If it is estimated that the mean density of matter in the universe is $10^{-29}$ gm/cm$^3$ and the mean mass of a galaxy is $8 \times 10^{10}$ solar masses, the number density of galaxies is approximately $6 \times 10^{-74}$

$cm^{-3}$. If the luminosity of a galaxy for all wavelengths is esti-
mated to be $8 \times 10^{10}$ times the solar luminosity, then $s = 2 \times 10^{-29}$ erg/sec $cm^3$. Using the value $H = 100$ km/sec Mpsc $= 3 \times 10^{-18}$ cm/sec cm, the estimated temperature of deep space is

$$T = 3.8° \text{ K} ,  \hspace{2cm} (\text{VIII.7})$$

which is in fair agreement with observations. (Considering the uncertainties in the estimate, Equation VIII.7, the agreement may be somewhat fortuitous.)

To complete the present discussion, Newtonian gravitational theory says that the gravitational energy of an expanding distribution of matter should be increasing. In particular, using a spherical model for the universe, the gravitational potential energy is given by

$$U = - G M^2/5r , \hspace{2cm} (\text{VIII.8})$$

where G is the universal gravitational constant, M is the mass of the universe, and r is the radius of the universe. The fractional change per unit time per unit volume is then given simply by

$$u^{-1}du/dt = r^{-1}dr/dt = H , \hspace{2cm} (\text{VIII.9})$$

where H is Hubble's constant. This gain in energy can be equated to the loss of energy by the deep space radiation resulting from the cosmological red shift if the energy density of radiation in deep space is set equal to the gravitional energy density—a not unreasonable assumption. Thus, the continued expansion of the universe, according to the present theory, is apparently forced by a gravitational coupling between radiation and matter, the energy being supplied by the radiation.

In conclusion, the features of the local environment which make life possible are duplicated throughout the universe. If the expanding universe theory is correct then it would appear that life exists only by virtue of the continued expansion. If the universe were to stop expanding, eventually thermal equilibrium would result and life would perish.

### Early Evolution of the Earth

It may be assumed that the earth condensed from a cloud of gas and dust encircling the sun. The angular momentum of the solar system indicates the necessity of planets evolving, since otherwise a sun containing all of the angular momentum of the solar system would tend to tear itself apart by centrifugal forces. As the gas and dust coalesced and cooled to form the earth certain elements and compounds condensed into solids and others into liquids. The solid and liquid phases have lower entropies than the gaseous phase, so that a supply of high utility energy was necessary to effect this entropy reduction. The energy came from the gravitational energy of the primordial gas cloud. As the cloud condensed, the gravitational energy was converted into thermal energy which was then radiated off into space.

The early ordering of the planet was quite extensive. The three phases—solids, liquids and gases—became spatially differentiated. Different solids and liquids became stratified according to densities, liquid iron settling into the earth's core. The plentiful existence of iron in the earth, despite its rarity in the universe, indicates that the earth formed under extreme entropy-reducing conditions. Elements in the early earth no longer remained haphazardly arranged, as in the original gas cloud, but segregated themselves into organised crystalline arrays. Large amounts of gravitational energy were expended to achieve this. As indicated in Table IV, the energy expended to form the earth was $2 \times 10^{39}$ ergs. By comparison the total solar energy dissipated on the earth in the 4.5 billion years of the earth's history has been about $150 \times 10^{39}$ ergs.

The temperature gradient in the earth indicates that the earth is still cooling off. Internal thermal energy of low entropy (i.e. high temperature) is conducted to the surface where it is radiated off into space as energy of high entropy (i.e. low temperature), permitting the possibility of entropy-reducing processes. Geologists sometimes attribute the formation of low entropic ores to percolating processes energized by the earth's heat. Thus, even today there are probably still ordering processes taking place in the earth which depend upon the gravitational energy that was once in the original gas cloud which gave rise to the earth.

Because the original condensation process, giving rise to the earth, was an entropy-reducing process, it is of interest to ask if life might have existed even before the sun turned on its nuclear furnace. There is ample evidence that life as simple crystal growth did exist, but there is no evidence that any advanced life existed by feeding on infall energy. The presence of a large potential for entropy reduction in a system containing the possibility of solids, the existence of numerous boundaries between phases, the existence of large thermal gradients, and the existence of turbulence in gases and liquids all suggest that advanced life might have existed on the earth prior to the irradiation of the earth by sunlight. Our present ignorance of detailed mechanisms that might have given rise to advanced life under such conditions may stem from our prior lack of a realization that life might, in fact, exist under such conditions.

After the solar system condensed out 4.5 billion years ago, the ordering of the earth's surface proceeded by virtue of irradiation by sunlight, as discussed in Chapter IV. At this juncture the earth was already well ordered, complex compounds were already separated, and life itself or precursors to life may have existed.

### The Entropy of the Earth Is Decreasing with Time

The entire evolution of a typical stellar system from its early stages as an interstellar cloud of gas and dust to its long period as a main sequence star with satellites to its final stage as a white dwarf is a continuous process of entropy reduction. Consequently, it is possible to postulate:

*A planet, revolving around a main sequence star, is evolving toward states of lower entropy.*

The rate of entropy reduction, once the planet has condensed out, may not be as great as during the period of the planet's original formation, yet ordering processes must continue. Even the moon and Mercury are probably still undergoing some slight entropy reduction in the outermost layers of atoms that are exposed to sunlight.

From the above postulate it may be concluded that the earth is presently evolving toward a state of lower entropy. If this is true, it should be verifiable by direct observations. The ordering that was produced by the original condensation of the earth from a cloud of gas and dust is of no interest here. It merely constitutes the initial condition from which further ordering is effected by irradiation by sunlight. The small rate of ordering produced by the earth's heat may be neglected here. To verify that the earth's surface is evolving toward a lower entropy state, lithospheric rocks may be compared with rocks that could not have evolved by virtue of solar radiation, such as meteorites, moon rocks and basaltic rocks welled up from the earth's mantle along the Midatlantic Ridge. Meteorites, moon rocks and deep basaltic rocks are characterized by fine-grained order. That is to say, if the rocks are inspected on a scale of about $10^{-2}$ cm, then compounds are clearly separated into different crystals. On such a fine scale basis there appears to be little to distinguish between the entropy of lithospheric rocks and the entropy of nonlithospheric rocks, except that lithospheric rocks appear to be composed of simpler compounds on the average. On a large scale basis, however, lithospheric rocks appear dramatically different from nonlithospheric rocks. On a large scale basis of $10^2$ cm or larger basaltic rocks, meteorites and moon rocks appear to be homogeneous mixtures of elements and compounds, whereas lithospheric rocks continue to appear as pure crystalline compounds. The lithosphere possesses vast deposits of limestone, iron oxide, salt, tar, sulfur, iron pyrites, coal, copper sulfide, quartz, clay, etc. It is clear that the surface of the earth started with complex compounds of fine-grained order which then became converted over geologic times to simpler compounds deposited with large scale order. Thus, it is evident by direct observation that the surface of the earth has, in fact, decreased its entropy over geologic time. Since the flux of sunlight has apparently not appreciably decreased in the last 4.5 billion years, the earth must still be undergoing a reduction in entropy.

## The Role of Life in Ordering the Earth's Surface

Life on the earth has probably played the major role in the ordering of the earth's surface subsequent to the original condensation of the earth. Autotrophs, such as green plants, remove high-entropy carbon dioxide, water and small amounts of other compounds and elements from the environment and convert them into low-entropy cellulose and other compounds which become deposited in the autotrophs. Green plants respire about 30 percent of the $CO_2$ and $H_2O$ originally deposited back into the environment. Detritus accumulates when autotrophs together with a smaller mass of heterotrophs die and are not immediately consumed. Herbivores eat green plants and detritus and respire $CO_2$ back into the environment. Carnivores, consuming herbivores, return another smaller portion of $CO_2$. Ground bacteria oxidize detritus. Forest, brush and grass fires further oxidize carbon fixed by autotrophs. The direct oxidation of carbon compounds exposed to the air accounts for some of the $CO_2$ returned to the atmosphere. A *carbon cycle* is, thus, established, the rate of reducing $CO_2$ and fixing carbon equalling the rate at which carbon is oxidized.

A small amount of detritus, however, does not get recycled, and it leads to permanent deposits of ordered compounds. Marine organisms dying and being deposited in a reducing environment produced deposits of oil, gas, sulfur and tar. Forests covered by silt have left large deposits of coal. Exoskeletons of microorganisms are responsible for the cliffs of Dover. Microorganisms may have selectively deposited out iron oxide in a shallow ocean bay to form the great Mesabe iron range. The intimate relationship between sulfide ores and limestone implies the possibility of deposition by marine organisms which were subsequently reduced. In fact, early microfossils are sometimes made visible by being outlined by microcrystals of sulfides (Barghoorn and Taylor, 1965). Coral reefs leave large deposits from a complex ecosystem. Peat bogs present a good example of current deposition of ordered compounds. This deposition of ordered compounds by life, while not important over a short period of years, becomes important over geologic times.

The greatest source of energy to effect ordering on the earth's surface has been direct sunlight. Since only autotrophs use large quantities of sunlight directly for ordering processes, it may be concluded that most of the ordering of the earth's lithosphere, since the condensation of the planet 4.5 billion years ago, has occurred through the agency of autotrophs. Sunlight also produces ordering processes indirectly in more purely geological environments. The percolation of rain water through soils and rocks with temperature and concentration gradients and subsequent evaporation and/or runoff to the ocean can produce selective crystallization of minerals. The dissipation of the earth's heat may also contribute to such ordering. It seems unlikely that any of these more purely geologic processes has produced the degree of ordering that has been produced by autotrophs and direct sunlight. However, the purely geological processes frequently mask the original contributions of living organisms.

The geologic dating of the strata in the earth's lithosphere since the Cambrian is traditionally done by identifying types of fossils present in the various layers. Life and the geology of the lithosphere are intimately commingled.

The total standing mass of living organisms (the biomass) and the detritus that is recycled, the ecomass, also represents a deposit of ordered compounds on the earth's surface. It appears that the amount of ecomass has been steadily increasing since the Cambrian, thereby contributing to the increasing order of the earth's surface.

Another way that the importance of life in the ordering of the earth's surface may be appreciated is by noting the amount of material that has passed through living organisms since Cambrian times. Assuming $4 \times 10^{16}$ gm of carbon are fixed each year on the earth at the present rate of photosynthesis, or one-fourth of this rate when averaged over the last billion years, the total mass of carbon cycled through living organisms over the last billion years is about $10^{25}$ gm. This may be compared with the mass of the earth's crust which is only about $2.4 \times 10^{25}$ gm, the lithosphere being only about one-tenth of the crust. Moreover carbon is not abundant on the earth, there being only about $10^{19}$ gm in the oceans and atmosphere and $10^{22}$ gm in the earth's crust. The total coal, oil, natural gas, etc., deposited by life in the lithosphere is perhaps $10^{19}$ gm. Elements such as calcium, phosphorous, sulfur and iron, which are frequently associated with life, can be preferentially concentrated and deposited by life. The resulting effects can be large because the flux of carbon and all such elements has been so large. It is, thus, possible to conclude that probably most of the ordering of the lithosphere was produced by life.

## A General Principle for the Evolution of Life

The previous discussion, indicating the role of life in the ordering processes occurring in the universe and on the earth in particular, leads to the postulation of a general principle for the evolution of life. This principle may then be applied to indicate some of the general trends of evolution in the past and to be expected in the future. In particular, the total mass of ordered compounds produced by life that is in permanent detritus or ore bodies and is in the standing ecomass increases over geologic or evolutionary times. Since the whole is the sum of the parts, it may then be postulated that the standing ecomass must also increase with time, it being assumed that each ecosystem on the earth is independent of all others, or, at most, only weakly coupled through such things as the $CO_2$ in the atmosphere. The fact that the ecomass cannot decrease with time (assuming no decrease in solar output) indicates a tendency, a pressure or a force driving an ecosystem toward a greater ecomass. This tendency toward a greater ecomass may be worded as a tendency toward a *maximum* ecomass, the failure to attain a maximum ecomass at any particular moment being attributed to secondary features, constraints or boundary conditions. The constraints might include such factors as the amount of incident solar energy averaged over a long period, the temperature of the earth averaged over a long period, the supply of available carbon, and the mean rainfall. The general principle postulated here is, thus:

*An ecosystem evolves toward a maximum ecomass.* (VIII.10)

Short time variations such as seasonal variations and times between generations of an organism are not being considered here. The time scale involved here is an evolutionary or geologic time scale of perhaps hundreds or thousands of years.

A number of important corollaries of this principle Equation VIII.10 may be immediately stated. Since the ecomass is usually proportional to the biomass (i.e. the mass of living organisms), a corollary is:

*An ecosystem evolves toward a maximum biomass.* (VIII.11)

Since again the whole is the sum of the parts, it may be said that, on the average considering all species:

*The biomass of a species will generally evolve toward a maximum biomass.* (VIII.12)

This corollary is, of course, weaker than the statement (VIII.10) so that the constraints and boundary conditions become more important. Again using the fact that the whole is a sum of the parts, the mass of a species being the sum of the masses of the individual organisms, it may be postulated that:

*An individual organism will generally behave so as to attain and maintain an internal state of maximum thermodynamic order.*

(VIII.13)

This statement is still weaker than Equation VIII.12, the constraints being still more important. Finally, noting that each individual species (and individual) is coupled through its thermodynamic processes to the ecosystem, it may be postulated that:

*An individual species (or organism) will generally behave so as to attain and sustain a maximum biomass for the ecosystem.*

(VIII.14)

This statement is the weakest and requires a complete detailed knowledge of an individual species role in the ecosystem.

The validity of the general principle of evolution, Equation VIII.10, and its corollaries postulated here will be explored in this and subsequent Chapters.

## Evolution, Genetic Change Plus Natural Selection

The fact that life evolved on the earth from primitive origins was recognized and documented over one hundred years ago by Darwin (1859). Darwin assumed that individual offspring of a species undergo each generation small deviations which make an individual more or less capable of surviving. An individual with altered traits that give rise to an enhancement of the chance to survive will have a greater chance of procreating and leaving descendents; consequently, such altered traits will be transmitted to succeeding generations. All subsequent theories of evolution have been similarly based upon the two fundamental ideas:

1. *the occurrence of inheritable deviations and*
2. *natural selection.*

(VIII.15)

(For background see Dobzhansky, 1951; Woolhouse, 1953; Lewontin, 1968; Broms, 1968; Drake, 1968; Dobzhansky *et al.*, 1967, 1968 and 1969.)

### Mechanisms for Producing Inheritable Deviations

Gregory Mendel (1865) discovered by experimental observation that certain traits in plants are inherited by rather simple rules, the Mendelian laws of inheritance. His results are explainable in terms of sexual reproduction where each sex has a set of paired chromosomes carrying the genetic information. Each parent contributes one member of each pair of chromosomes to the offspring, so that the offspring again has the same number of paired chromosomes as the parents, except that the genetic information (genes) on each member of the paired chromosomes may be different. It is not always easy to decide the phenotype to be expected when two different genes operate on corresponding chromosomes. Sometimes one is dominant and the other recessive, as originally discovered by Mendel, but sometimes they may act in a mutualistic manner. The problem becomes further complicated when more than one pair of chromosomes carry genes affecting a single trait. Since each chromosome can carry a large number of genes, even a few pairs of chromosomes per individual can give rise to a huge number of new variants from an initially heterogeneous population (Elandt-Johnson, 1971).

The Mendelian laws (or, more precisely, chromosomal genetics), while certainly operative, are not sufficient, since there must eventually be changes in the genes themselves to give rise to the different genes found in heterogeneous populations. Because of the large number of chromosomal types possible the rate of change of genes need not be very great to account for the rate of observed deviations. It was thus assumed for many years that inheritable deviations could be adequately accounted for by gene changes, called *mutations*, plus chromosomal inheritance.

There has been considerable research to try to discover the mechanisms whereby genes may be altered. It is obviously not sufficient to simply apply the word "spontaneous" to mutations and ignore the actual mechanisms whereby they could arise. Muller (1941) demonstrated in the late 1920's that genes could be altered by strong ionizing radiation. Potent irradiation could also produce new sequences of old genes by such mechanisms as crossovers of chromosome pairs. The new genes produced by such artificial means are almost always harmful to the individual, essentially none being useful. It was assumed for many years that gene changes, or mutations, occur randomly in nature due to background radiation or thermally induced disruptions. But natural background radiation is not sufficient to account for the observed rate of mutations, and thermal disruptions would be much too rare to account for the observed rate. The fairly recent verification of the fact that double-stranded DNA (the material from which genes are made [Drake, 1970]) is able to repair itself upon being damaged further detracts from the possibility that most mutations can be explained as randomly induced events.

It is also clear that inheritable deviations arising from mutations plus chromosomal mixing is not applicable to a large number of species that do not reproduce sexually. For example, bacteria show a remarkable rate of adaptibility of inheritable traits. It was once assumed that this ability resulted from a much higher mutation rate, taking into account the fact that each bacterium is also a germ cell. It is now known that the explanation is the direct transfer of genetic material from one bacterium to another, which circumvents the usual slow process in higher organisms for establishing new genetic types. Genetic material can be transferred between bacterium (Burns, 1972) by (1) taking up naked DNA from the medium left by other individuals, (2) conjugation where material is passed through a male pilus to a female cell (Curtis, 1969), and (3) transduction where a viral-like particle, or episome, invades a cell and attaches itself to the host DNA to be replicated along with the host DNA (Campbell, 1972). Thus, without having to reproduce a bacterium can mutate. A favorable mutable trait can be thus transferred directly to the same generation without the long delay necessitated by many generations.

Paramecia are also observed to mix their genetic material between individuals by conjugation. In the absence of other individuals with which to conjugate they eventually undergo an internal rearrangement called endomixus. Thus, paramecia, which do not reproduce sexually, apparently have some sort of ability to alter genetic information similar to chromosomal mixing. The direct transfer of genetic information and the possible subsequent alteration of the individual is comparable to the direct transfer that occurs in bacteria.

It is becoming increasingly apparent that life is separated into two broad general classes: eukaryotes, which are organisms comprised of cells which possess a central nucleus with genetic information and organelles with some limited genetic information; and prokaryotes, which are organisms comprised of a cell which has no nucleus, the genetic information being distributed throughout the cytoplasm. Eukaryotes, which constitute the more highly evolved class, apparently arose as a symbiotic group of prokaryotes inclosed within a single cellular membrane (Margulis, 1971). It may be envisioned that the first eukaryotes arose when one prokaryote invaded another as a parasite, a phenomenon that can still be seen today. Then by evolutionary adaptation through natural selection it may be assumed that the prokaryotes became symbiotic and lost their ability to live independent of each other. Mitochondria and chloroplasts are organelles that appear to be such symbiotically evolved prokaryotes, since they are small; they still retain a degree of autonomy; and they carry much of the genetic information for their own reproduction. The evolution of the main nucleus of an eukaryote could not be independent of the evolution of its included semiautonomous organelles, and vice versa. The mechanisms for the generation and propagation of inheritable deviations of such collections of semiautonomous individuals is necessarily complicated.

Viruses have been known for over a hundred years. The work of Pasteur with the rabies virus established many of the characteristics features of a virus. A virus is now known to invade a cell and to introduce information which makes the cell generate more virus particles. It was long suspected and finally discovered that in some cases the virus particle attaches itself directly to a chromosome in the nucleus of the host cell (Campbell, 1972). For good or bad, such an appended piece of information may be regarded as an additional gene. It has now been verified that in at least one case such an appended viral gene can be transmitted to subsequent generations of the host cell. It has been speculated for some time that viral infections may provide the mechanism for most mutations of cellular organisms. It now appears that such a possibility will have to be taken seriously. If true, then the mutability of the virus would determine the ability of life to produce inheritable deviations.

It is apparent from the discussion in the present section that as man learns more and more about the mechanisms for the production of inheritable deviations he finds the situation increasingly more difficult to analyze in any simple way. It will be sufficient here to merely note that inheritable deviations do occur and not to inquire any further into the possible mechanisms for their occurrence.

## An Evolutionary Hierarchy of Living Things

It appears that living things fall into a gross hierarchical structure that recapitulates evolutionary development as follows:

1. Simple compounds such as amino acids crystallizing and polymerizing which collect into organized arrays of polypeptides and proteins.
2. Collections of proteins and other compounds forming structures identifiable as virus-like particles.
3. Collections of virus-like particles living symbiotically within a single protective membrane which are identifiable as prokaryotes.
4. Collections of symbiotically adapted prokaryotes living within a single cell membrane identifiable as eukaryotic cells.
5. Collections of eukaryotic cells identifiable as macroscopic life forms as ordinarily seen by man.
6. Collections of macroscopic organisms interacting to form an ecosystem.

From the observation of life forms around us today we can identify organisms coexisting which belong to practically all levels on the phylogenetic tree. Where man lives, frogs still live; where frogs live, mollusks still live; where mollusks live, protozoa still live, where protozoa live, algae still live; where algae

live, bacteria still live; where eukaryotes live, prokaryotes still
live; etc. It may, thus, be conjectured that the lowest two categories in the above hierarchy should also exist today. It should
be possible to find today the crystallization and/or polymerization of amino acids in the environment without the presence of
advanced life forms, and it should be possible to find free-living
virus-like particles which are not parasitic on any advanced life
forms. The discovery of such free-living particles would be extremely interesting, but unfortunately they cannot be easily
found due to their extreme smallness and due to their very small
effect upon the environment. Infectious viruses are known simply
because of their drastic effects on large organisms. The apparent indetectibility of free-living virus-like particles in no way
detracts, however, from their possible role in the evolution of
life or in their possible present-day role in distributing genetic
information.

## Natural Selection and Maximum Biomass

The prediction of the end product of evolution rests upon the
concept of natural selection. While the time necessary to achieve
certain evolutionary results may depend upon the rate that inheritable deviations arise, the physical properties of the life
forms that finally evolve depend only upon natural selection.
Darwin appreciated this fact in his enunciation of the principle
of the survival of the fittest. If the measure of fitness were survival, then the principle would reduce to a tautology of little
value. But it is not necessary to use survival as a measure of
fitness; an inspection of the environment can frequently lead to
a prediction of the type of organism that would be most fit. For
example, in a desert the fittest will evolve mechanisms for conserving water. In the ocean the fittest will tend to evolve mechanisms for excreting salt. In the air the fittest will tend to become
very small, develop a low density, evolve wings, or evolve glid-
ing planes. The fittest treetop-foraging ungulate will evolve long
front legs and a long neck, etc.

The principle of survival of the fittest can also be used to
predict the nature of a species that involves other species. For
example, a spider that preys on crawling insects may not evolve

a web. A bird that must feed on hard-shelled nuts will evolve a short strong beak to crack the nuts. Darwin, noting the dimensions of the flower of a night flowering plant, correctly predicted the existence of a moth with the appropriate length proboscis. Such detailed predictions are of great interest, but the purpose of the present section is to explore the more general and theoretical consequences of the general principle postulated above, Equation VIII.11.

EVOLUTION TOWARD HIGHER ECOLOGICAL EFFICIENCIES. *Ecological efficiency* is defined here as the ratio of the mass of a living system to the rate of expenditure of energy necessary to maintain the system. This measure can vary from zero to infinity. When comparing efficiencies the energy involved should be of the same utility. The rate that energy is available to an ecosystem averaged over a sufficient number of years is a constant equal to the rate that sunlight is available. If the system tends toward a maximum biomass, Equation VIII.11 it will then also tend toward a maximum ecological efficiency. It has been estimated by Ehrensvärd (1962) that the biomass of the earth from the Cambrian to the present has increased almost twentyfold. Such estimates are very uncertain, being taken from the fossil record, but it may be safely assumed that there has been at least some increase in the biomass since the Cambrian. Since the amount of sunlight has remained fixed, the ecological efficiency of the earth has increased.

When life has been removed from an area it becomes re-established in a sequence of stages which is referred to as *succession*. Succession recapitulates past evolution much like ontogeny recapitulates phylogeny and the evolution of species. As life becomes re-established on a denuded area it may be observed that the biomass increases. The biomass continues to increase with each seral stage, mature communities having the greatest ecological efficiency.

Considering the corollary, Equation VIII.12, evolution will tend to favor the more ecologically efficient of two varieties or species competing for the same niche. For example, a green plant that can put out a greater leaf area per unit of energy absorbed is more ecologically efficient, and it will be able to outshade its competitors and will be able to obtain a larger share of the available sunlight. A similar argument holds for all food gathering. The species that can put the most foragers into the field for the amount of food consumed will tend to leave less for competitors.

If two varieties or species compete with active hostility where they attempt to actively destroy each other, the ecologically more efficient, being able to put more combatants into the field per amount of food consumed, will be better able to eliminate competition.

A more efficient species will be able to support a large number of individuals per unit area of the earth's surface. Consequently, a more efficient species will tend to physically displace or crowd out the less efficient species. In the struggle for territory or living space the more ecologically efficient species will win out.

VARIABLE ENVIRONMENTS EVOLVE SPECIES WITH MAXIMAL REPRODUCTION RATES. If an ecosystem evolves toward a maximum biomass, Equation VIII.11, then any periodic or random depletion in the supply of food or sunlight which causes a decrease in a population should be followed, upon a renewal of the food supply or sunlight, by the most rapid recovery of the population, in order to maximize the time average biomass. Since the mass of living organisms is primarily altered by a change in the number of individuals, an increase in the food supply means reproduction at a maximum rate. Since food supplies do frequently change markedly with the season or with the weather, it may be concluded that species dependent upon such variable food supplies will evolve a capacity to reproduce at a maximal rate. This conclusion derived from the general principle postulated above may be verified by considering some specific arguments.

As indicated in the previous subsection, two species with the same ecological efficiencies will be able to survive together on the same steady food supply indefinitely without their relative numbers changing with time (Ayala, 1971). However, if the two species have different reproduction rates and they are presented with a sudden increase in the food supply, they will both increase in numbers to take advantage of the increased amount of food, and the species that can increase more rapidly will end up with a proportionately larger population. This increased ratio for the more rapidly reproducing species will not decrease with time and will not decrease even when there is a subsequent reduction in the food supply. Any repeated fluctuation in the food supply either periodic or random will then eventually yield the survival of the species that can reproduce at the faster rate and an extinction of the other species. This argument helps to explain why the observed capacities of most species in nature to reproduce at rates far beyond the rates necessary to fulfill ordinary steady state requirements occur.

It should be noted that it is the rate of reproduction that is involved here and not the final population; a fast rate of reproduction does not imply a runaway population. Species populations are generally limited by factors in addition to food, so that the food supply upon which they feed is not destroyed by overpopulation with a subsequent extinction of the species. Actually, few species are ever in a position to threaten their own food supply by overpopulation, most food supplies being fixed independent of the species population, such as sunlight for green plants.

PROLIFERATION OF SPECIES AND INCREASING BIOMASS. Evolution has proceeded from organisms of simple structure to a great variety of organisms with complicated structures. This proliferation of species and types may be accounted for by the general principle that the biomass of an ecosystem tends toward a maximum. The greater the range of environments in which the totality of organisms can survive, the greater the total biomass that can be maintained. The greater the range of environments in which organisms can live, the greater will be the variety of specialized structures and organs to equip the organisms for survival. It may be assumed that species specialized to fit a narrow range of environmental conditions will be generally more ecologically efficient than those equipped for a large range of conditions. There is, thus, a tendency for different species to evolve to fit different niches rather than for one species to evolve to fit many niches, since evolution is in the direction of higher ecological efficiency. For example, it would undoubtedly be very inefficient to have an animal equipped both as a ruminant and as a flying insectivore, as compared with two different species, cows and bats. The proliferation of species to fit a wider range of environments and narrow niches permits a greater total biomass to be sustained.

It is also possible, using Darwin's principle of the survival of the fittest, to account in detail for specialized structures needed by species to exploit new environments. For example, Darwin's study of the finches on the Galapagos Island showed the relationship between the variety of species according to bill type and the variety of foods available to them (Lack, 1947).

Succession of an ecosystem from an original denuded area, which recapitulates the evolutionary process, shows an increasing number of species and an increasing biomass with time. It has been frequently noted that there is a relationship between the biomass and the number of species, arctic tundra having few species and little biomass while tropical rain forests have many species and a large biomass. To be properly checked, however, the same climatic conditions should prevail; thus, the biomass on two islands with the same climates but with different numbers of species should be compared. The island with the larger number of species should have the greater biomass.

Noting the proliferation of species, one might wonder why some primitive species continue to survive. The reason is clear: primitive species occupy niches in habitats where life has existed the longest; therefore, species in such habitats have had ample opportunity to evolve into the most efficient organisms for their particular niche. Species structure and form remain invariant over vast periods of time, not because of any inability to produce inheritable deviants, but because natural selection always reverts all deviations back to the most efficient form, the original form. (It should be noted, however, that for carbon-based life evolution toward actual ideal forms is blocked by the limited capacity of carbon-based life to produce viable deviants with sufficiently different characteristics. No such restriction applies to machine life.)

Species that do succeed in evolving new traits are forced to seek out new environments, niches or habitats. The earliest deviants that exploit a new environment will not have had an opportunity to evolve the most efficient traits possible. Thus, generally speaking, the most highly evolved species should be found in the most inhospitable environments, should be the most ecologically inefficient, and should, therefore, be subject to the most rapid evolutionary changes.

EVOLUTION OF LARGER INDIVIDUAL ORGANISMS. It is observed that organisms have an average metabolic rate T (energy consumed per unit time per unit body mass) that is roughly proportional to the reciprocal one-third power of the mass M,

$$T = K M^{-1/3} , \qquad \text{(VIII.16)}$$

where K is roughly a constant for all species equal to about 1.4 kcal/gm$^{2/3}$day (Odum, 1959). In other words, the energy flux into and out of an organism is roughly proportional to the surface area, while the mass maintained is proportional to the volume. Since the metabolic rate is just the reciprocal of the ecological efficiency for the individual, the evolution of larger individuals will tend to allow for an increased biomass. Thus, evolution is in the direction of larger individual organisms.

This conclusion is borne out by observations. Proceeding up the phylogenetic tree one encounters larger and larger individuals. While it may not be possible to follow the process precisely in geologic time, it is still significant that the present-day whale is the largest animal that has ever existed on the earth.

Even though larger individual organisms are apparently favored evolutionarily, the precise mechanisms that come into play are not clear. However, a few advantages may be noted: Many of the interactions that an organism has with its environment are proportional to its surface area. The small surface to volume ratio (the specific surface) of a large organism provides a relative isolation from all surface effects for the large organism as compared with a small organism. A large organism will be able to remain comfortable in extreme heat and extreme cold. The effects of surface-living parasites relative to the mass is small for large organisms. Collisions and abrasions are relatively insignificant for the large organism. The velocity that a motile organism can attain is usually roughly proportional to its length, or to the one-third power of the mass, $M^{1/3}$. Thus, a large organism will be better able to escape small predators and/or will be better able to capture small prey.

EVOLUTION OF SEEDS AND SEED DISPERSAL. As discussed in Chapter IV there is no advantage for adult autotrophs to evolve motility, since sunlight, $CO_2$ and $H_2O$ are provided more or less uniformly. However, autotrophs do frequently exhibit a considerable ability to disperse seeds and spores to favorable sites where the adults again become sessile.

A habitat subject to wide seasonal and random changes will support more biomass at one time than another. In order to maximize the time average biomass, species and individuals that die off at one time must be replaced as rapidly as possible when conditions improve. This can be achieved by spores and seeds remaining dormant in the habitat during unfavorable conditions giving rise to new individuals when conditions improve. Seeds and spores are hardly alive according to the definition of life presented in Chapter III; consequently, they are essentially independent of their environment, yet they have the potential for life and can emerge from their dormant state to grow into active adult autotrophs. The evolution of seeds and spores with dormant capabilities, thus, provides for a maximum biomass.

A site may, due to changing circumstances, become suitable for the growth of a particular species of plant where the plant has never grown before. The biomass of the biosphere would be increased if the plant were to grow in the new site. Seeds and spores being generally quite small and requiring little or no externally supplied sustenance are transported by wind, flowing water and motile heterotrophs. The transport of seeds may be envisioned as a migration of the species involved. The evolution of seed dispersal tends to provide a maximum biomass.

## Life Span and Evolution

The principle of natural selection assumes that individual organisms must die. The differential death rate between the fit and the unfit then gives rise to a new generation with more fit genes than unfit genes. Death is, therefore, an integral and necessary feature of the evolutionary process. An important question is then: Is the age at death selected by evolutionary processes, or is the age at death merely a necessary end result of living? It is clear that a high death rate, which permits more generations per unit time, will permit a high rate of evolution. If species have to adapt by rapid genetic changes, they might be expected to evolve a short life span or early death. It does not appear, however, that death is a product of evolution. Instead it appears that death is a necessary end result of living. In particular, it appears that:

*Each organism dies after it has processed a certain fixed amount of energy per unit body mass.*                         (VIII.17)

It will be seen below that this fixed amount of energy for mammals is about 800 kilocalories per gram fresh body mass. This idea was originally introduced by Rubner (1908) who by comparing the metabolic rates and life spans of different species arrived at the figure of 700 kcal/gm. This relationship between the life span and energy consumed, Equation VIII.17, provides another example of the fruitfulness of applying thermodynamics (and choosing energy as a variable) when studying biological and ecological systems. The matter is worth pursuing further here.

If Equation VIII.17 is true then the life span L of an individual should be related to its mean metabolic rate (per gram body mass) T by

$$L = E_o/T,\qquad\qquad \text{(VIII.18)}$$

where $E_o$ (equal to about 800 kcal/gm for mammals [see below]) is the fixed amount of energy consumed per unit body mass per lifetime. This relationship can be checked in a number of ways (*see* the review by Bourlière, 1958). Cold-blooded animals have metabolic rates which increase with temperature; therefore, cold-blooded animals living in different temperature environments should have life spans which vary inversely with the temperature. This is found to be the case. At one end of the scale a frozen amoeba, which can always be revived by thawing, can be regarded as having an infinite life span, $L \to \infty$. Experiments with fruit flies (Pearl, 1928; Alpatov and Pearl, 1929) obtained a mean life of 29 days at 28° C and 70 days at 18° C.

Experiments with a water flea by Terao and Tanaka yielded a maximum life span of 4.8 days at 33° C, 6.5 days at 27° C, 9.3 days at 21° C, and 14 days at 15° C. Similarly, for *Daphnia magna* (MacArthur and Baillie, 1929) the mean life span was found to be 26 days at 29° C, 42 days at 18° C, 88 days at 10° C, and 108 days at 8° C. Similar results were obtained for planarian worms (Abeloos, 1930). It has also been noted that insects develop more rapidly in the tropics than in temperate zones (e.g. Fountaine, 1938). Frogs and fence lizards are observed to have much shorter life spans in southern United States than in the northern part (Oliver, 1955). Fish when raised in cooler water live longer.

Warm-blooded animals usually increase their metabolic rates at low temperatures (as well as at elevated temperatures), so that experimentally rats raised at low temperatures do not live as long as rats raised at normal temperatures. Johnson *et al.* (1963) and Kibler *et al.* (1961) found that rats raised at 9° C had a 40 percent elevation in metabolism and had a maximum life span of 700 days, while rats raised at 28° C had a maximum life span of 950 days. According to Equation VIII.18 this would imply that $1.4 \times 700$ should equal $1 \times 950$, or 980 compared with 950, a satisfactory agreement with theory.

The metabolic rate of both warm- and cold-blooded animals can be reduced by reducing the caloric intake. Thus, the life span of rats has been considerably extended in the laboratory by withholding calories (McCay, 1942; Berg and Harmison, 1957). The inverse effect has been noted in pigs and humans. When they are fed more than adequately they grow bigger and mature younger thereby resulting in a reduction in the life span as expected by the present theory. It has been noted that the age of pubescence of girls in boarding schools has decreased almost three years in the last century and a half, apparently as a result of improved nutrition since they also increased in height.

An animal that is more active and does more mechanical work should use up its allotted calories faster and should die earlier. Such is observed to be the case in honey bees (Ribbands, 1953). Summer bees work hard foraging and live only about a month, while fall bees do little or no work and live till the following spring. The actuarial statistics on man fails to show that hard laborers die younger than those in sedentary professions. Evidently, hard labor in humans does not represent a sufficiently large increase in the net consumption of energy over a man's entire life to allow the effect to be observed readily. Other determiners of the lifetime effective metabolic rate may also be more important in humans, such as obesity, hypertension or the amount of sleep. Since it is observed that fat men die younger than skinny men, it must be presumed here that the effective lifetime metabolic rate of fat men is greater than skinny men. Experiments exercising rats have not produced enough difference in the net rate of energy consumed to allow the expected shortening of the life span to be observed.

Another way to check the theory, Equation VIII.18, is to note that the metabolic rate varies as $M^{-1/3}$, Equation VIII.16, so that the life span should vary as $M^{1/3}$; thus,

$$L = K'M^{1/3} , \qquad (VIII.19)$$

where $K' = E_o/K$ is a constant with the approximate value of $1 \text{ gm}^{-1/3}\text{year}$. This is plotted in Figure 30 on a log-log scale. Data is also presented in Table VI where maximum observed life spans are used as estimates of the physiological life span (Alman and Dittmer, 1964). It might be argued that Figure 30 indicates a spurious relationship, since a tree cannot actually live hundreds of years. The supporting wood contains no living tissue, and the individual cells in a tree live no longer than one or two years. And at the other end of the scale, a bacterium never dies (unless killed by extraneous means); it merely divides. To make the theory a little more plausible, the case of the sequoia tree might be neglected, and the life span of one-celled organisms can be taken as simply the maximum time between cell division. The relationship, Equation VIII.19 appears to be sufficiently established by Figure 30 with its extremely wide range in the variables and Table VI to inquire into possible reasons for discrepancies from an exact fit. The wet body mass introduces a spurious effect; for example, flying animals and desert animals are relatively dehydrated compared with a jellyfish. Dry body mass would be better. To circumvent the problem of nonliving

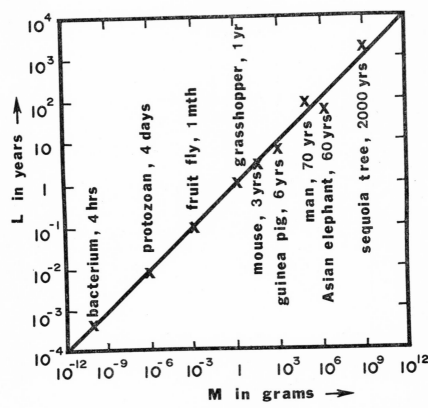

Figure 30. The approximate relationship between the life span L in years and the mass of an organism in grams, Equation VIII.19, with a few species indicated.

fat storage in some species, the mass of nitrogen might be better. The mass of nonliving material such as skeletons should also be omitted.

Species may also differ appreciably in their modes of energy expenditure. Some birds and bats are capable of periods of extremely low metabolism (Bartholomew *et al.*, 1957; Reeder and Cowles, 1951). Animals that sporulate, hibernate or remain dormant for long periods should have longer life spans than those that remain active. Animals that sleep more should live longer. Herbivores tend to sleep little, while carnivores sleep for long periods each day. Thus, in Tables VI and VII the cow, horse, goat, European rabbit, guinea pig, golden hamster and mouse should have life spans shorter than predicted, while man, rhesus monkey, dog and cat should have life spans greater than predicted; this is seen to be true in every instance. Cold-blooded animals have quiescent stages of very low metabolism, so that reptiles, amphibians and fish might be expected to have longer life spans than mammals or birds of comparable dry mass.

TABLE VI. COMPARISON OF THE THEORETICAL LIFE SPAN
GIVEN BY $K'M^{1/3}$ AND THE OBSERVED MAXIMUM AGE AT
DEATH $L_m$ WHERE $K' = 10yr/kgm^{1/3}$

| Species | Common Name | Mass (kgm) | $K'M^{1/3}(yr)$ | $L_m(yr)$ |
|---|---|---|---|---|
| Bos taurus | cow | 600 | 84 | 30 |
| Equus caballus | horse | 450 | 77 | 50 |
| Homo sapiens | man | 80 | 43 | 70 |
| Polyodon spathula | paddlefish | 74 | 42 | 24 |
| Capra hircus | goat | 65 | 40 | 18 |
| Acipenser fulvescens | lake sturgeon | 50 | 37 | 15 |
| Cyprimis carpio | carp | 38 | 34 | 47 |
| Esox lucius | northern pike | 28 | 30 | 24 |
| Gadus morhua | atlantic cod | 25 | 29 | 16 |
| Ictalurus punctatus | channel catfish | 24 | 29 | 13 |
| Salmo trutta | brown trout | 19 | 26 | 10 |
| Lepisostera osseus | longnose gar | 18 | 26 | 30 |
| Canis familiaris | dog | 16 | 25 | 34 |
| Microterus salmoides | large-mouth black bass | 10 | 22 | 16 |
| Macaca mulatta | rhesus monkey | 8.0 | 20 | 29 |
| Coregonus clupeaformis | North American whitefish | 5.0 | 17 | 26 |
| Melanogrammus aeglefinus | haddock | 2.7 | 14 | 15 |
| Felis catus | cat | 2.5 | 14 | 21 |
| Gallus domesticus | chicken | 2.5 | 14 | 30 |
| Pomoxis annularis | white crappie | 0.9 | 9.6 | 9 |
| Cavia porcellus | guinea pig | 0.7 | 8.9 | >6 |
| Mesocricetus auratus | golden hamster | 0.1 | 4.6 | 1.8 |
| Mus musculus | house mouse | 0.03 | 3.1 | >3 |

The postulate, Equation VIII.17, may also be checked directly where both T and L are known approximately for certain species. Because the total energy consumed throughout the lifetime of an individual depends upon its behavior, the appropriate value for the mean lifetime metabolic rate T, which must be estimated from quiescent values, is not always apparent. For mammals (with the exception of the bat) the resting metabolic rate is generally proportional to the lifetime average metabolic rate, so that Equation VIII.18 may be roughly checked for mam-

TABLE VII.  COMPARISON OF THE PHYSIOLOGIC LIFE SPAN $L_o$ ESTIMATED FROM THE METABOLIC RATE $T_o$ IN MILLILITERS $O_2$ PER KILOGRAM FRESH BODY MASS PER HOUR, $L_o = K^*/T_o$ IN YEARS, WITH MAXIMUM OBSERVED LIFE SPAN $L_m$ IN YEARS

| Species | Common name | $T_o$ | $L_o$ | $L_m$ |
|---|---|---|---|---|
| *Mammals* | | $K^* = 10^4 \ yr \ ml \ (O_2)/kgm \ hr$ | | |
| Elephas maximus | Asiatic elephant | 155 | 65 | 57 |
| Dasypus sp. | armadillo | 201 | 50 | 18 |
| Sus scrofa | swine | 220 | 46 | 27 |
| Homo sapiens | man | 220 | 46 | 70 |
| Equus caballus | horse | 250 | 40 | 50 |
| Ovis aries | sheep | 340 | 29 | 20 |
| Bos taurus | cow | 390 | 26 | 30 |
| Procyon cancrivorus | crab-eating raccoon | 395 | 25 | 14 |
| Ornithorhynchus sp. | platypus | 460 | 22 | 14 |
| Phoca vitulina | harbor seal | 540 | 19 | 19 |
| Canis familiaris | dog | 580 | 17 | 34 |
| Oryctolagus cuniculus | European rabbit | 700 | 14 | >13 |
| Felis catus | cat | 710 | 14 | 21 |
| Cavia sp. | guinea pig | 815 | 12 | >6 |
| Rattus sp. | rat | 2000 | 5.0 | 3.3 |
| Mesocricetus auratus | golden hamster | 2900 | 3.4 | 1.8 |
| Mus musculus | house mouse | 3500 | 2.9 | >3.0 |
| *Flying birds and bats* | | $K^* = 5 \times 10^4 \ yr \ ml \ (O_2)/kgm \ hr$ | | |
| Corvus corax | raven | 940 | 53 | 69 |
| Myotis lucifugus | little brown bat | 1500 | 33 | 23 |
| Serinus canarius | canary | 2900 | 17 | 24 |
| *Reptiles* | | $K^* = 10^3 \ yr \ ml \ (O_2)/kgm \ hr$ | | |
| Alligator mississipiensis | American alligator | 8.2 | 122 | >56 |
| Malaclemys terrapin centrata | southern diamondback terrapin | 35 | 29 | >21 |
| Anguis fragilis | slowworm | 40 | 25 | 32 |
| Natrix sp. | European water snake | 121 | 8.3 | 7 |

TABLE VII. (CONTINUED)

| *Pisces (exclusive of lung fish)* | | $K^* = 2.5 \times 10^8$ yr ml $(O_2)/kgm$ hr | | |
|---|---|---|---|---|
| Carassius auratus | goldfish | 85 | 29 | 30 |
| Cyprinus carpio | carp | 100 | 25 | 47 |
| Esox lucius | northern pike | 102 | 25 | 24 |
| Anguilla sp. | European fresh water eel | 128 | 20 | 15 |
| Salmo trutta | brown trout | 226 | 11 | 10 |
| Scomber sp. | Atlantic mackerel | 726 | 3.5 | 15 |

mals by comparing estimated physiologic life spans with observed resting metabolic rates. Table VII presents material from the *Biology Data Book* (Alman and Dittmer, 1964). The oxygen consumption of a resting adult animal in milliliters (at standard pressure and temperature) per kilogram fresh body mass per hour $T_o$ has been used to yield the expected physiologic life span $L_o$ measured in years using the formula

$$L_o = K^*/T_o , \qquad\qquad (VIII.20)$$

where for the best fit $K^*$ has been chosen as $10^5$ years milliliters of oxygen per kilogram fresh body mass. This theoretical physiologic life span is then compared with the maximum age observed at death. Besides maximum observed life spans being crude estimates of the physiologic life spans, there are inaccuracies in the measurement of oxygen consumption. These measurements are extremely sensitive to temperature, to age of the subject, sex, prior state and a host of other variables. Assuming 5 kilocalories per liter of oxygen consumed and assuming the lifetime average metabolic rate is 60 percent higher than the resting rate, the value of $K^*$ and Equations VIII.20 and VIII.18 yield the resulting lifetime caloric consumption $E_o = 800$ kilocalories per gram fresh body mass for mammals. Due to dehydration $K^*$ has been increased by a factor of five for birds and bats. Because of long periods of quiescence, $K^*$ has been reduced by a factor of ten for reptiles. To fit the case for fish $K^*$ has been reduced by a factor of four.

All carbon-based life appears to be governed by the same mechanism causing aging and death, but it is not clear what this mechanism is. The wear and tear theory of aging is compatible with the present observations if it is assumed that the rate of wear and tear is proportional to the rate of energy consumption. The aging process might be viewed as the depletion of some vital compound which is never renewed after the birth of the cell, the rate of depletion being proportional to the rate of energy consumption. Since single-celled organisms may be viewed as living forever if the identity of an individual is assumed to remain intact even after a cell division, the mechanism of aging is necessarily linked to cell division or to replication. To a rough first approximation it may be assumed that two cells are born and one cell dies when a cell divides. While a death factor appears to be passed on by some tissue cells, successive cell divisions being limited to some set number of generations (a function of age of the multicelled organism) after which no further cell divisions occur, this is not generally the case. For example, the original navel orange tree with thousands of successive grafts indicates no limit to the number of generations. In general, it appears that a cell after consuming its alloted amount of energy must die either by ceasing to function or by dividing. The life span of multicelled animals is, thus, apparently determined by the life span of the essential cells that do no divide (e.g. nerve cells). (Gelfant and Smith [1972] present a model for aging in proliferating tissue.) Once enough such essential cells have died the total organism experiences some type of failure and dies.

It would be interesting to explore the age at death of humans with extreme metabolic pathologies, such as progeria, since their life spans should also be determined by Equation VIII.18. Further details and speculations may be found in the literature (e.g. Strehler, 1962; Curtis, 1966; Andrew, 1971).

## Life when Evolution Ends

If the evolutionary process were to continue indefinitely, then the thermodynamic order of the earth's surface would have to increase indefinitely. But this is impossible; eventually a steady state of minimum entropy would have to become established (assuming no diminution in the solar energy available). The question is then whether or not life would continue to exist even in this final steady state of maximum order. Life, being essentially a cyclic steady state process, does not imply that evolution is necessary for its existence. Thus, life might possibly continue to exist in a nonevolving biosphere. However, from the previous sections it is apparent that such life as remained would tend to be less living (according to the definition of *living* presented in Chapter III), since the direction of evolution is toward greater ecological efficiencies, or less energy consumed per body mass.

# CHAPTER IX

# EVOLUTION IN
# THE FUTURE

E VENTUALLY UNDER the action of sunlight, neglecting machines
and the direct utilization of nuclear energy, the earth would
evolve to a steady state of maximum thermodynamic order at
which time all further evolutionary processes would cease. The
earth today is apparently still far from such a state of maximum
order. Only a small fraction of the available incident solar energy
is currently being used to achieve and maintain a steady state of
thermodynamic order, while in principle a much larger fraction
could be used (*see* Chapter V). Evolution could proceed far into
the future much as it has done in the past. If carbon-based life
were all that had to be considered, this would undoubtedly come
to pass. The advent of the machine, however, has produced an
abrupt change in the course of evolution. The future evolution
of the earth's surface toward a state of higher thermodynamic
order will be extremely different from the evolution of the past,
since it will involve the future evolution of machines. No matter
how different the future may appear, the same general principles
governing evolution as discussed in the previous Chapter will still
be valid in the future, since, according to the definition of life
presented in Chapter III, machines are also living.

## The Concept of Ecomass Extended to Machines

In Chapter VIII the concept of the biomass of an ecosystem
played a major role in the discussion of evolution. Namely, eco-

systems tend to evolve toward a maximum time average biomass. But precisely what is biomass? Is it the mass of living material in the ecosystem? Is the shell of a clam, the hair on a bear, the wood inside of a tree, or the exoskeleton of a beetle part of the biomass. Is the water in an organism part of the biomass? Is blood plasma, amniotic fluid, bone, cartilage or other tissue without cells included in the biomass? Are the gizzard stones in a chicken part of the biomass of the chicken? Should amoebas that are frozen for the winter, and therefore nonliving during the winter, be included in the biomass during the winter? Actually, these questions present no real difficulty, because the concept of a maximum time average biomass arose simply as a corollary of the more general principle, Equation VIII.10, that the ecomass tends toward a maximum, or the surface of the earth evolves toward a state of lower entropy. Except for the water in organisms and the gizzard stones of a chicken, the examples cited above represent a steady state deposit of low entropic compounds which are part of the recycling ecomass.

The distinction between biomass and ecomass, while not especially important in the case of carbon-based life (since they are essentially proportional), becomes more important when considering machine life. A considerable quantity of machines and tools are nonliving, but they affect the ecosystem. A screwdriver or a monkey wrench do not consume energy and are, therefore, nonliving; yet in the hands of a man they can help to create order in the ecosystem. A wheel on an automobile is nonliving, but it is essential to the performance of the whole automobile which is living when the engine is running. Few of the component parts of a machine are living, in contrast to carbon-based life where practically every portion of an organism contains living cells. A lathe which is driven by a belt functions as a complex tool since it consumes work instead of some other form of energy, and it may be regarded, therefore, as nonliving. An automobile with its engine turned off is certainly not living, yet it is still important, since sometimes it comes to life and functions in the ecosystem. Whether or not certain individual items are living or nonliving is not actually important here. The important question is whether or not they contribute to the ecomass. The items mentioned above are composed primarily of iron. Crystalline iron is in a lower entropy state than the entropy of iron ions in sea water or the iron in iron oxide ore from which the iron originally came. This iron, in principle, is being recycled. The items mentioned above are, therefore, all parts of the ecomass. In order to put iron mass and carbon mass on the same scale as they function in the ecosphere, one atom of iron should be equated to two-thirds atoms of carbon; or the mass of iron should be multiplied by the factor one-sixth, as explained at the beginning of Chapter VI.

## Prehistoric Precursors of Modern Machines

In terms of the 4½ billion years of evolution of carbon-based life on the earth the advent of modern machines has been amazingly abrupt. Attempting to discover how it all came about, it may be noted that there were three essential precursors to modern machines: tools, fire and intelligence.

EVOLUTION OF TOOLS. Tools and implements have been used for many millions of years by many widely different species of animals; for example, birds' nests of twigs, gizzard stones, rocks thrown by apes, mud daubers' nests of mud, wasps' nests of paper, lobsters' ear stones, hermit crabs' shells, rocks on which gulls drop clams, twigs on bagworms, thorns used by birds to probe for insects, and straws used by chimpanzees to eat ants (van Lawick-Goodall, 1971). A primitive hominid (Leaky, 1968) in Kenya 15 million years ago must have used tools to kill his large prey. Australopithicus, with his small teeth and large prey undoubtedly used crude tools of bone, wood and stone (Washburn, 1959). After the brain evolved to three times the size of Australopithicus, true man appeared half a million years ago (if not earlier) with the tools he inherited from preman. Slowly the complexity, variety and usefulness of man's tools increased. Over the last ten thousand years there has been an explosion in the evolution of tools, there being a countless variety today.

EVOLUTION OF FIRE. The first evidence of controlled fire is about 500,000 years ago by the slightly primitive precursor of modern man, *Pithecanthropus pekinensis* (Hawkes and Woolley, 1963). This remarkable event occurred very recently in terms of the whole time scale of 4.5 billion years of the earth's evolution, being only a $10^{-4}$ part. Pithecanthropus and his cave fires were probably not particularly important quantitatively in their early ecosystem, except possibly in the increased rate of occurrence of forest, brush and grass fires.

The advent of fire implies an entirely different sort of ecological role for the fire user. Early fire users probably burned wood that had already fallen from the trees; it is unlikely that they cropped living trees. The early fire user adopted the simultaneous roles of wood decomposer and carnivore. If it is assumed, for example, that a 50 kilogram fire user burned 5 kilograms of wood each day to yield 20,000 kcal/day, then the fire user increased his caloric consumption by a factor of about ten over the prefire user. A sudden increase of about an order of magnitude in the energy consumption of a species represents a marked change in the ecological role of the species.

It might be reasoned that the fire user, playing the double role of a carnivore and a wood decomposer, enjoyed double security by filling two niches simultaneously. There probably has not been sufficient time to determine whether or not this double role did, in fact, confer any evolutionary selective advantage on man.

The situation might also be viewed as the sudden introduction of a new species of organism, *controlled fire*, which has a very small biomass (the mass of the hot gases in the flame) and very high metabolic rate. The new organism lived symbiotically with the fire user. Since the biomass of a flame is very small, the ecological efficiency of fire is much less than that of other wood decomposers (although the ecological efficiency of bacteria is also very low). In free competition with bacteria, it would appear that bacteria should replace fire. However, it may be noted that fire produced naturally by lightning has always managed to exist despite its lack of ecological efficiency. Eventually machines, replacing carbon-based life, may replace forests, brush, and grass and thereby replace such fires.

Even if the unique event of the introduction of controlled fire cannot be fitted into the more ordinary ideas of ecology, it must be assumed that fire yielded some benefit to the fire user. Fire was probably introduced originally for space heating, having been introduced almost half a million years ago during a cold epoch. Even today most of the energy consumed by man in the home is devoted to space heating. A warm-blooded animal, such as a human, normally maintains its internal temperature at a constant high level. A human being maintains an internal temperature of about 98.4° F (36.9° C) and ordinarily dies if the temperature varies as much as 8° F (or 4° C) from this value. When placed in a cold environment a warm-blooded animal will consume more calories in order to maintain its internal temperature. For example, the metabolic rate of rats may be increased 40 percent by putting them in a cold environment of 9° C (Johnson *et al.*, 1963). The advantage of space heating (as well as clothes) to man in a cold environment is, thus, clear. He need not consume as many calories to maintain his internal state of constant temperature. Thus, a smaller portion of the total fixed amount of energy consumed in a man's lifetime need be consumed merely to maintain an internal state of constant temperature. The energy saved can then be used to maintain and create order in other ways.

The energy required by primitive man to collect fuel to burn was undoubtedly much less than the energy conserved by the burning of such fuel for space heating. Such space heating could also allow man to move out into less hospitable environments, such as mountain tops, ice age tundra, and polar regions.

It is a curious fact that warm-blooded animals have lower ecological efficiencies than cold-blooded animals. When the environmental temperature drops the temperature of cold-blooded animals also drops with a corresponding decrease in energy consumption, while warm-blooded animals when the environmental temperature drops continue to maintain their high temperature by wasting energy. In addition, as a heat engine an organism should ideally prefer a cold source (the ambient temperature of the environment) to be as low as possible, in order to perform at the greatest efficiency. Yet warm-blooded animals function best at some fixed high temperature (around 70° F [or 21° C] for humans), thereby operating at a lower than ideal efficiency. Evidently, the high temperatures are required for a high rate of activity. Insects must, apparently, also be warm to function rapidly (e.g. Heinrich, 1973).

It might be asked how warm-blooded animals, with their lower ecological efficiencies, ever evolved from their cold-blooded ancestors. The only way it could occur was for warm-blooded animals to move to new environments, habitats or niches. It may be conjectured that eventually warm-blooded animals will evolve to again become cold blooded while remaining in their present niches.

Controlled fire was probably also important for reasons other than space heating, such as: providing short-ranged vision at night to lengthen the working day and to see predators (especially in a cave), providing techniques for food preservation by smoking and drying, providing a way to kill harmful parasites in food, providing a method for frightening or burning would-be predators, providing techniques for fire hardening wooden spears and smelting and reducing ores, providing the technique of *lighting* fish and game at night, providing techniques for communication over long distances, providing a method for breaking down plant tissue to make plant products more digestible to a meat eater, and providing methods for making pottery. The ability to manipulate the environment in numerous ways was, thus, placed in the hands of the fire user.

EVOLUTION OF INTELLIGENCE. Another necessary precursor to modern machines was the evolution of a brain large enough to invent modern machines. While the Neanderthal, as well as other varieties of early man, may have been more intelligent than modern man (some had larger brains than modern man (Leaky, 1960), human culture had not evolved to the point where such early intelligence could be used to create machines. Similarly, the porpoise, which, according to anatomical evidence, is about as intelligent as modern man, and certain whales (Cousteau and Diole, 1972), which may be much more intelligent than modern man, have no tools or culture.

It seems strange that despite all of the apparent advantages given to early man—the controlled use of fire, the ability to fashion and use tools, two free prehensile hands, a large brain and a social structure—he still seems to have been hard pressed to survive. How is this fact known today? It is known quite simply by the evolution of a bigger brain which modern man has inherited. The larger brain is not a gift, an attribute which somehow escaped the process of natural selection; it is a result of brute survival. The stupid perished, the intelligent lived and reproduced. The niche that man filled was obviously fraught with great peril, as it must be even today. Only an exaggerated intelligence, along with all of the other apparent advantages such as tools and fire, was barely able to keep man alive. Since the principle of parsimony operates in evolution, only the most ecologically efficient survive. A superfluously large brain could not evolve; it had to be absolutely necessary for survival.

It is difficult today to see just exactly why the niche that man filled required the evolution of such a large intelligence. Man is certainly not inferior in stamina; the Hopi Indians use to run deer to the ground, and a man in condition can cover one hundred miles in twenty-four hours. Man is not inferior in strength. Man is generally superior in eyesight. The most primitive club or spear makes man in a hunting pack superior to any predator and, in fact, made man the predator of the largest animals on the earth. It is perhaps only in the sense of smell that man has apparently developed deficiencies. While it might be argued that man developed his large intelligence by virtue of competition with his own kind, this cannot be true, since all species evolve to fit the requirements of their niche and the niche does not include the organisms themselves. If the niche which man filled required a smaller intelligence, then the more successful organisms would be the ones that did not have to support an unnecessarily large brain. Evolution produced the least intelligent individuals necessary to perform their ecological role.

### Evolution of Machines

Man during prehistorical times and historical times developed a wide variety of devices to enchance his own physical effort. He developed levers, the wheel, pulleys and numerous clever devices of string, bone, wood, iron and bronze. When man invented the boat with a sail he had available sources of mechanical power in the wind, tides, river currents and ocean currents. In Roman times man first started using water power with the water wheel. In about 1000 A.D. man harnessed the mechanical power of wind to turn windmills. In about 1660 the second Marquis of Worcester built the first practical device for utilizing the energy of steam (Usher, 1929a). Savery improved on the device in 1700. These devices could only pump water. The final step that brought together the necessary ingredients for modern machines—tools, fire and intelligence—occurred when Newcomen invented his steam engine in 1712. His device could, in principle, be used to deliver mechanical power at any permanent installation. This introduced the modern era of living machines.

In the previous Chapter a number of general tendencies in the direction of evolution of carbon-based life were explored. Since the totality of machines is also themodynamically alive, the evolution of machines is subject to precisely the same evolutionary principles. In particular, the ecomass, including machines, tends toward a maximum, Equation VIII.10, and the two fundamental ideas, Equation VIII.15,

1. the occurrence of inheritable deviations and
2. natural selection,

still remain valid for machines.

It may be noted that machines reproduce. Whatever the mechanism for their duplication and propagation, the fact remains that they are duplicated and they are propagated. Each year sees new Chevrolets on the road, the old ones slowly disappearing from the highways. New automobiles appear to be much the same as the old ones; they are still identifiable as automobiles with most of the detailed features the same. It is, thus, apparent that the information for their production, or more precisely *reproduction*, survives from generation to generation of automobiles. The new generation of automobiles is a product of this genetic information which is passed down from car manufacturers of one generation of automobiles to car manufacturers of the next generation of automobiles. The genetic information and its inheritance remains outside of the bodies of the machines, yet the end result is precisely the same as for carbon-based life where the information for the manufacture of the next generation is contained within the bodies of the previous generation. Carbon-based life retains a fantastic redundancy of genetic information, each cell of a muticelled animal containing precisely the same total information.

Machines are *rationally* devised, which means that their characteristics are determined by the minds of designers and inventors. Using their knowledge of engineering, science and economics, they are responsible for the alterations found in new machines. Successful alterations that yield the manufacturer greater profits are kept in future generations of machines. Thus, such alterations are, in principle, inheritable (strictly speaking, in the line of designers and inventors, rather than in the machines themselves).

Natural selection operates for machines. The uneconomical machine does not get sold, and, therefore, it does not subsequently get reproduced. The unsuccessful machine, yielding no profit to the manufacturer, will disappear from the surface of the earth, or will become *extinct*. While the mechanisms for the inheritance and alteration of genetic information is drastically different for machines as compared with carbon-based life, the law of natural selection operates in much the same way, i.e. by trial and error.

EVOLUTION TOWARD A MAXIMUM ECOMASS. The general principle that the earth tends toward a time average maximum ecomass implies that machines tend toward a maximum mass. This may be checked historically. For example, considering iron as a measure of the biomass of machines, the United States produced $2.0 \times 10^{12}$ gm of pig iron in 1860 and $8.3 \times 10^{13}$ gm in 1970, a 41-fold increase in this period. During the same time the human population increased only six times while the carbon-based ecomass remained roughly constant. The world production of iron from ore in 1968 was $3.7 \times 10^{14}$ gm with a doubling time of about 11 years. Photosynthesis fixes about $4 \times 10^{16}$ gm of carbon per year. The carbon equivalent of iron is one-sixth the mass of iron, or the world production of iron in 1968 had a carbon equivalent of $0.6 \times 10^{14}$ gm. At the present doubling rate in about 100 years the rate at which iron will be reduced will exceed the rate that carbon is reduced by photosynthesis.

Atom for atom aluminum amounts to about 10 percent of the iron production in 1968. Its doubling rate is about 5 years, which means that more aluminum will be produced in sixty-five years than carbon will be fixed by photosynthesis.

The material of which machines are made does not appear to be recycled as rapidly as carbon compounds, so that machines are rapidly accumulating a large mass which should soon surpass the biomass of carbon-based life.

EVOLUTION TOWARD INCREASED EFFICIENCY. Not only is there an evolutionary tendency toward greater ecological efficiency (mass supported to energy consumed), there is also a tendency toward increased efficiency in the performance of specific ecological roles. Thus, machines whose purpose is to deliver mechanical power while consuming chemical or thermal energy should evolve greater efficiencies (work out to energy in). The first engine of Newcomen was undoubtedly less than 1 percent efficient in converting chemical energy to mechanical work. After fifty years of experience in building such engines efficiencies increased up to about 1.5 percent (Usher, 1929b). Watt's improvements were able to double the performance of the steam engine bringing the efficiency up to 3 or 4 percent by 1800. By 1850 the performance of stationary steam engines increased to almost 15 percent. By 1900 the efficiency rose to about 18 percent. While steam turbines can now yield efficiencies up to about 40 percent, internal combustion engines, gas turbines and fuel cells have pushed the efficiency for the conversion of chemical energy to work to almost 70 percent. Little additional improvement can be expected due to natural thermodynamic limitations.

The efficiency of a horse or a man in delivering mechanical work (work done to the chemical energy in the food consumed) over a twenty-four hour period is no more than about 5 to 10 percent on an actual working basis. This efficiency exceeds that of the Newcomen engine; so why was the Newcomen engine able to survive? In terms of the mechanical work delivered, man, the horse and the Newcomen engine all played precisely the same ecological role, but in terms of the food supplied they each belonged in a different niche. Man required the most expensive high caloric food, the horse was able to survive more cheaply on primarily cellulose, while the Newcomen engine survived on the cheapest food of all, low grade coal. The Newcomen engine competed for coal only with those demanding coal for space heating. The engine was built at the mine where there were no transportation costs which were high in those days. The Newcomen engine with its low efficiency survived because it was exploiting a new niche, and with time it was replaced by more efficient devices. Heat engines operating on fossil fuels are essentially detritus feeders where the detritus cannot be assimilated by any other organisms to produce useful work.

It may also be verified that the ecological efficiency of machines has been increasing with time as it should to obey the general evolutionary principle. Taking the pig iron production in the United States as a rough measure of the ecomass of machines and dividing by the total energy consumption to get an approximate measure of the ecological efficiency of machines, it is found that this ratio increased by about seven times in the century from 1850 to 1950 (U.S. Census, 1960). Since the doubling time for the increase in energy consumption is about 25 years, while the doubling time for iron production is about 11 years in 1970, the ecological efficiency will continue to increase in the future.

EVOLUTION OF MAXIMAL REPRODUCTION RATES. It is clear that there is an economic advantage to increase as rapidly as possible the production of machines for which there is a sudden increase in demand. For example, such an increase in demand might arise for harvesters after a bumper wheat crop. Manufacturers try to gear their reproduction of machines to fit the market; they try to reproduce machines as fast as possible when the demand arises. There is, thus, a tendency for machines to have a maximal reproduction rate under changing conditions just like carbon-based life.

PROLIFERATION OF SPECIES. The specific roles required of machines have continued to increase so that the proliferation of machine types or species has been fantastically great over the last 250 years. An adequate taxonomy of machines has not as yet been devised, even though it would probably be desirable. Generally the nomenclature of machine types has been limited to their immediate function. A "lawn mower" can be anything from a device that one pushes to a self-propelled device of 4 or 5 horsepower that consumes gasoline and one sits on. Machines, like carbon-based life, is evolving in the direction of an increasing number of species and varieties.

There appears to be no particular advantage to larger machines, such as was true for carbon-based life, so that machines do not seem to be evolving toward larger sizes (unless the role demands it).

EVOLUTION OF MOTILITY. As explained in Chapter IV there is evolutionary pressure for heterotrophs to evolve motility, motility yielding advantages in gathering food and eliminating wastes. Machines, being heterotrophs, have also evolved motile species such as cargo ships, passenger ships, automobiles, trailer trucks, pick-up trucks, delivery trucks, motor bikes, motorcycles, earth movers, bulldozers, submarines, freight trains, passenger trains, airplanes, subways, els, busses, helicopters, jets, electric golf carts, hover crafts, spaceships and farm tractors.

EVOLUTION OF INCREASED LIFE SPANS. As with carbon-based life there is a tendency for machines to evolve in the direction of a longer life span. In terms of human economics it merely means that the replacement costs for a machine per unit time decrease when the machine lasts longer. Automobiles, for example, appeared to have a life span of about three years in the 1920's, about five years by the late 1930's, about six or seven years in the 1950's, and today they appear to have a life span of almost ten years.

### Machines Will Make Carbon-based Life Extinct

By considering the general principles of evolution it becomes apparent that machines will ultimately win in the struggle for survival between machines and carbon-based life. Some carbon-based life has already been replaced by machines. The point of no return was probably reached in about 1830 when the steam engine became more efficient than man or horse even if the steam engine were to burn the same expensive fuel as used by man or by the horse. From that point onward it might have been predicted that machines would eventually dominate the earth and that carbon-based life would become extinct.

The words "extinct" or "extinction" are used here in a somewhat more technical sense than is frequently employed. In a very narrow or absolute sense a carbon-based species becomes extinct when all individuals of that species have perished together with all of the genetic information necessary for the replication of the species. Ecophysics, being more of a quantitative science, is not particularly concerned with the concept of absolute extinction. Ecophysics is more concerned with the quantitative role of a species in the ecosphere; and when the quantitative role has essentially vanished with no hope of ever being quantitatively resurrected, then the species is regarded as being extinct. The whooping crane, the large whales hunted to the verge of absolute extinction are within the present context *extinct*. Within the present section the extinction of carbon-based life is taken to mean the eventual reduction of carbon-based life to an ecological role of vanishing importance from which it will never recover.

HISTORICAL REPLACEMENT OF CARBON-BASED LIFE BY MACHINES. In 1800 a horse or ox was generally used to transport goods and men from place to place in the United States. By 1875 the steam locomotive had largely replaced the horse and ox for long hauls. The horse was still the predominant prime mover, however, for transporting goods over short distances until about 1915. By 1930 practically all horses had been replaced for short hauls by automobiles, trucks and tractors. In 1875 most of the heavy farm labor was still done by men and horses, and rarely by steam-powered harvesters and tractors. By 1950 essential all heavy farm labor was taken over by tractors. The machine had replaced most of the horses on farms by 1930. At one time tens of thousands of Missouri mules were a source of pride to Missourians; today they have all vanished except for a few individual mules kept for sentimental reasons.

Rubber trees have been largely replaced by factories manufacturing superior synthetic rubber products. Indigo was once a big crop which has been entirely replaced by synthetics. Various species of bacteria have been replaced in the manufacture of methyl alcohol, acetic acid, urea, nitrates, etc. Rayon, nylon, polyesters, Spandex,® polyethylene and other synthetic fibers produced by machines have replaced a large number of organisms such as sheep, cotton plants, goats, hemp plants, sisal plants and silk worms. Wood, bone, gut, leather and ivory have been replaced by plastics, steel and aluminum. Linseed oil from flax, tung oil from the tung nut, and other natural drying oils have been replaced by emulsified acrylic polymers, polyvinyl alcohols, and other synthetic resins of superior quality.

Machines have taken up room. They have crowded out carbon-based life in order to have room for themselves. Carbon-based life cannot resist a bulldozer. The area of industrial plants, roads, railroads, parking lots, switching yards, chemical dumps, warehouses, etc., is constantly increasing at the expense of carbon-based life. Railroads and roads together with fences have helped to make the larger migrating and far ranging animals essentially extinct in America, such as the buffalo (i.e. bison), the elk, the wolf, the puma and the grizzly. Roads also interfere with the movements of some species over short distances as indicated by the large death toll of certain animals on the American highways. Dams on many waterways have blocked fish from their traditional spawning grounds which has resulted in the destruction of many varieties of fish, if not whole species.

Machines and devices such as the gun, which lives briefly as it disgorges its projectile, have made extinct or essentially extinct numerous species. Whether or not man *controlled* these devices is not the issue here. Rather, it is simply noted that the machines and devices did the killing by contributing thermodynamically in a major way. The passenger pigeon, the whooping crane, the heath hen on Martha's Vineyard, the bald eagle, the laughing crane, the egret and numerous other species of birds have been made completely extinct or essentially extinct (i.e. the population is less than $10^{-3}$ times the original population). Practically all large North American animals have been made essentially extinct by the gun: the wolf, bison, grizzly bear, puma, elk, wolverine, antelope, alligator, moose, mountain goat, porcupine, mountain sheep, coyote, bobcat, beaver and brown bear. Certain species of whale have been made essentially extinct and some species (Small, 1971) will probably become absolutely extinct through the slaughter of whales by machines. Whales have little chance against the superior speed of ships driven by steam, the use of sonar, the use of aircraft, and the lethality of fired harpoons with exploding charges.

Machines have produced a large amount of destruction of carbon-based life merely incidentally to the activity of the machines. For example, at one time sulfide ore smelters permitted sulfur to escape into the atmosphere, which then killed the vegetation over a wide region, such as at Copper Basin at Copperhill, Tennessee. Strip mining for coal has left thousands of square miles of the United States denuded of carbon-based life. Dredging for sand and gravel has similarly led to the destruction of carbon-based life. Abandoned mines generally leave a pile of tailings on which little carbon-based life can grow for a hundred or more years. Dredging, hydraulic mining, placer mining and hard rock mining for gold in California have left large areas still deficient of their original carbon-based life.

The excretia of machines, i.e. industrial wastes, have produced some spectacular kills in streams, rivers and lakes from time to time. Many ocean bays and beaches are polluted and dead due to industrial wastes. Oil spills from wrecked super tankers and ruptured offshore drilling equipment have produced some major catastrophes for the carbon-based life along some of the beaches of the world. The thermal effluent of some power plants into small streams has changed the ecology of the stream, killing certain species. Machine-produced and machine-spread DDT continues to destroy vast numbers of birds. Rachel Carson (1967) and others (e.g., Kneese and Bower, 1972) have indicated the extent of the problem. The continued increase in the pollution rate of air, earth and water by machines means a steadily increasing death rate for carbon-based life. Machines are generally immune to such pollution.

Machines have wrought havoc to the carbon-based ecology of South Vietnam. Airplanes have repeatedly spread herbicides over large areas to defoliate and kill trees and rice crops (e.g. Herbicide Assessment Commission of the AAAS, 1970). Many large mangrove forests will never recover. Lifeless craters left by bombs now cover an important fraction of the formerly productive land area. Huge bulldozers have stripped the top soil off of large areas in order to destroy the carbon-based life.

THE GREATER EVOLUTIONARY ADAPTABILITY OF MACHINES. Machines are *rational* organisms in the sense that they are planned and devised by an intelligence before they are ever put into practical operation. An inventor can anticipate the performance of his invention. Not only does the knowledge of how to manufacture a machine get passed down from one machine generation to the next, but there is also a continuous evaluation of the performance of machines which results in changes in subsequent generations. Thus, machines may be said to inherit acquired characteristics, i.e. inherit any observed characteristics deemed valuable by the inventors no matter how the characteristics might have been acquired. In addition, designers and inventors can anticipate the success of their devices and may, thus, give machines favorable characteristics which have never been acquired.

The only limit on the evolutionary rate for machines is the rate at which new inventions or modifications can be generated. The limitation on the rate of the evolution of machines today is, thus, set only by the inventiveness of human minds. The inventing process may at times appear to be slow to an engineering firm that wants a device today and has to wait ten years until it is developed, but compared with the evolutionary rates of carbon-based life this is no time at all. It might take carbon-based life 100,000 years, 1,000,000 years or 10,000,000 years to achieve comparable results. As more and better electronic computers are brought into play the rate that successful modifications can be made will increase. Eventually, it may be anticipated that electronic brains will do most of the inventing and designing. Since electronic brains are machines themselves, they can help design themselves, which implies at least an exponential growth in the evolutionary rate of machines.

The greater evolutionary rate of machines implies that once a machine has evolved to the point where it outperforms a carbon-based species it will continue to improve at a faster rate and the carbon-based species will never be able to catch up. The greater evolutionary rate of machines implies a rapid proliferation of types which will be able to outperform all of the different carbon-based species with their inherent inferiorities. The greater evolutionary rate implies that machines will be able to fill new niches long before carbon-based life could ever evolve to the point of utilizing these niches. A spectacular example of the exploitation of a new niche by machines is in outer space where communication satellites, revolving about the earth, feed directly on sunlight.

In contrast to the rational evolution of machines, carbon-based life has evolved agonizingly slowly over hundreds of millions of years by an excessively wasteful process of *irrational* inheritable changes plus the destruction of countless individuals to permit the fittest to be chosen. There has not been sufficient time by such a slow process to evolve carbon-based organisms that are ideally equipped to play their roles. There is hardly any specific task or job performed by a carbon-based organism that cannot be immediately improved upon, in principle, by a clever inventor. For example, the advantages of the wheel are numerous and obvious, yet no carbon-based organism has ever been able to evolve wheels. No nocturnal species has ever evolved infrared vision. Apart from the primitive use of electric current by electric eels and fish, no carbon-based species has evolved to utilize the extremely valuable and versatile electric current. Basic errors in metabolic pathways and in structure in primitive organisms are propagated on to more advanced organisms, the primitive errors being then irremedial. For example, man could use another pair of hands and arms, lungs that permit a unidirectional flow of air, another backbone, etc. Every advance in carbon-based life is built upon a heterogeneous collection of traits representing uncorrected errors, mistakes and successes. Each new generation may be better adapted than the one before, but each generation still falls far short of the ideal. There is no way for the evolution of carbon-based life to ever proceed within any reasonable period of time (less than $10^{11}$ years) to any sort of ideal end product as might be readily envisioned by an inventor.

THE GREATER EFFICIENCY OF MACHINES. Machines appear to have ecological efficiencies comparable to that of carbon-based life today. For example, comparing a 60 kg human with a 1500 kg automobile (with respective equivalent carbon masses of 12 kg and 250 kg), the human consumes 2400 kcal/day while the automobile consumes about 600 gallons of gasoline per year (at

$3 \times 10^4$ kcal/gal). Both of their ecological efficiencies are then about the same, $5 \times 10^{-3}$ gm day/cal. However, in the future as the ecological efficiency of machines continues to increase while that of carbon-based life remains the same, the ecological efficiency of machines will surpass that of carbon-based life.

The total accumulated value of the mechanical work delivered by a horse plus the value of the carcass of the horse (or the resale value) divided by the total accumulated cost to keep the horse plus the initial cost of the horse gives a measure of the overall economic efficiency of a horse in its role. Similarly, the total accumulated value of the mechanical work delivered by a tractor plus the resale value of the tractor divided by the total accumulated cost to keep the tractor running plus the initial cost of the tractor gives a measure of the overall economic efficiency of a tractor playing the same role. In actual practice the tractor has been found to be the more efficient in the role than the horse and has, thus, replaced the horse.

It takes energy to order raw compounds into finished products. A production efficiency may then be defined as the mass of the finished product to the energy required to produce it. Machines have proven to have superior production efficiencies to carbon-based life for numerous compounds such as dyes, urea, methyl alcohol, acetic acid, rubber, plastics and for all reduced metals (carbon-based life having zero production efficiency in this case).

GREATER REPRODUCTION RATE OF MACHINES. Once a successful machine or device has been invented, mass production can usually be achieved in a short time. Because of the nature of the mass production of machines, a rapid adjustment of the number of machines produced (i.e. reproduced) can be achieved to fit changing circumstances. A sudden doubling of arable land could result in a doubling of the number of tractors reproduced over a few months or years. In contrast the reproduction rates that can be achieved by carbon-based organisms that might compete with the tractor are much slower, it requiring probably five to ten years to double a horse population and twenty-five to fifty years to double a human population. As indicated in the last Chapter, with everything else equivalent, the variety that can reproduce at the faster rate to take advantage of sudden favorable changes in the environment will replace the slower reproducing variety. If machines and carbon-based life are equivalent in every way except in the rate of reproduction, the faster reproducing machines should replace carbon-based life.

MORE ENERGY AVAILABLE FOR MACHINES. Most of the energy sources that can be currently envisioned (*see* Chapter V) are not directly available to carbon-based life. Water power, coal, tides, volcanic heat, nuclear fission, etc., are all available only to machines. Carbon-based life could never evolve in any reasonable period of time to utilize these sources of energy directly. In the case of solar energy, carbon-based life does use the energy directly, but carbon-based life uses only a small fraction that could be consumed by a heat engine using mirrors to focus the sun's rays on a boiler. Since the amount of biomass that can be supported is proportional to the high utility energy available, it appears that the potential biomass of machines is greater than that of carbon-based life. The biomass of machines should exceed that of carbon-based life within one hundred years.

SUPERIOR SENSORS AND EFFECTORS OF MACHINES. In the competition for survival the variety finally selected is frequently the one that has a superior ability to detect the nature of its environment (e.g. the ability to detect a predator by an improved sense of smell or sight) and that has developed superior effector organs for taking the appropriate action (e.g. longer tusks to impale a would-be predator). It may, thus, be assumed that the possibilities available to machines for superior sensors and effectors should give machines the evolutionary advantage over carbon-based life.

To merely list a few of the superior features of machines the following may be noted: Machines can lift and transport weights that are many hundreds of times greater than those that can be handled by the largest animal. Machines can operate in radioactive environments completely destructive to carbon-based life. Machines can communicate by electromagnetic waves over millions of miles of space. Machines can *see* individual atoms. Machines can *see* the details of distant objects such as the billion light year distant quasars. Machines can *feel* the heat of a match a hundred miles away. Machines can multiply two 3-digit numbers in a microsecond. Machines can *feel* the weight of a gnat's wing. Machines can obliterate a large area halfway around the world using a rocket with a nuclear warhead. Machines can *know* when $10^{-9}$ seconds has elapsed. Machines can *sense* distances of $10^{-8}$ cm. Machines can change matter into energy. Machines can project bullets to kill at distances of hundreds of feet. Machines can produce temperatures of 0.1 degree absolute. Machines can produce temperatures of 10,000,000 degrees absolute. Machines can make plutonium from uranium and can transmute other elements.

## Machines Where Carbon-based Life Never Lived

Quite apart from the competition between machines and carbon-based life in niches presently occupied by carbon-based life, evolutionary principles reveal an extremely important future for machines. Machines can live in extreme ranges of temperature, pressure and chemical environments. It is conceivable that machines, utilizing solar energy, could survive in the Sahara Desert without a source of water in the environment. To carbon-based life the oceans present an even larger desert than the Sahara, yet, in principle, machine organisms floating on the water and using solar energy could also survive on the oceans.

It is in space, where carbon-based life has never evolved any self sufficiency and never will, that machines will develop their full potential. The numerous advantages to space have already been pointed out in Chapter V. It will probably be machines that will travel to the other planets. Already machines have landed on Venus and Mars, and there are currently three in orbit around Mars. It will probably be machines that will first set up self-reproducing colonies on the other bodies in our solar system and that will ultimately become established on worlds about other suns. Earthbound carbon-based life will probably never share this expansion of machine life among the stars.

## Machines Will Make Man Extinct

When one reflects upon extinct species such as the dinosaurs, the sabertoothed tiger, or the woolly rhinoceros, one is prone to ask what species might replace an extinct man. The great success of man among carbon-based species together with the extremely slow evolutionary rate of carbon-based life indicates that no carbon-based species is likely to replace man in the near future. The fact that machines will make all carbon-based life extinct also means that machines will replace man. While it may not be possible to predict the final ultimate extinction of every individual member of the species *Homo sapien,* it can still be readily predicted that man's ecological role will be reduced to one of vanishing importance from which he will never recover. Man will become essentially extinct if not absolutely extinct.

THE HISTORICAL SUCCESS OF MACHINES OVER MAN. Ever since the early days of the industrial revolution, almost two hundred years ago, various tasks originally performed by men have been taken over by machines. The net replacement of men by machines has been so drastic that today man spends almost no time doing the tasks that formerly occupied all of his working time, and essentially 100 percent of the working time spent by man today is on tasks completely unknown two hundred years ago (e.g. Jerome, 1934).

MAN'S FAILURE TO STRUGGLE AGAINST THE MACHINE. During the industrial revolution men were sometimes thrown out of their jobs because they were replaced by machines. Since this deprived them of a livelihood, they saw the machine as competitors and a direct threat to their survival. In their frustration they sometimes rioted and destroyed a few machines (Mantoux, 1961). The struggle for brute survival was clearer then. Now man prides himself in being more sophisticated. He no longer runs around smashing machines merely because he has been displaced from his job by a new machine. Society with its unemployment benefits, featherbedding, union funds and special relief now tides a man over until he can find another way to make a livelihood. The assumption now is that for every new machine introduced there is a new job created. Man no longer struggles against the machines. And the bravest and most courageous attempts during the industrial revolution appear today to have been weak and stupid.

The ballad of "John Henry," purportedly based upon a real competition between a steam-hammered drill and John Henry who used his muscles to hammer a rock drill, is one of the most poignant folk songs that has ever survived. John Henry, according to the song, won the competition against the steam-hammered drill, but died in the process. It is implicit that he not only died due to his great exertions, but also due to the fact that he knew that he would never again be the greatest "steel driving man." While countless thousands of men have undoubtedly been similarly psychologically crushed and beaten after attempting the same foolish sort of competition with a machine, it is a humiliation generally hidden and never mentioned. Such a *foolish* competition is rarely made public. Each individual man replaced by a machine has the personal choices: (1) try to destroy the machines, (2) try to compete with the machines (like John Henry), or (3) prudently slink off to fill a niche that machines have not as yet invaded. Naturally, the third choice is inevitably taken. It has the stamp of public and social approval, and only a fool would fight the inevitable.

The problem that remains, however, is that eventually man will run out of niches free of machines. Eventually all men will then be back in precisely the same position as the man in the industrial revolution who, when he was displaced by a machine, faced actual starvation. When this point is finally reached it will be realized that man never put up any real struggle against the machine. It was hopeless in any case.

While most men have never seriously assumed that machines will ever take over, there have been occasional individuals who felt that machines might win out (e.g. Chase, 1929; Ellul, 1964). There have also been quite a few more individuals that recognized that automation (i.e. the continued encroachment by machines) presents some serious problems (e.g. Scott and Bolz, 1969 and 1970; Marcson, 1970; MacBride,1967; Markham, 1963), but they generally felt that the problems were not insoluble or that the problems would generally solve themselves given sufficient time. There has never been sufficient awareness among a sufficient number of men of the true depth of the problem facing mankind, in order for any action to have been taken against the machines (neglecting, for the moment, the futility of any action).

Even if everyone in the world were to read this book and

realized the danger of man's extinction by the machine, it would still have no effect upon the outcome. Each individual man is forced to follow the cruel dictates of the narrow day-to-day situation in which he finds himself. Men are individually prisoners of the superiority of machines. A man who works on an assembly line aids and abets the machines used. If he wished to turn back the tide of events and wrecked the machines, or if he decided that his work was supporting the machines and quit, the result would be the same. In either case he would be without a job, and consequently without a means of livelihood. Another individual not quite so *insane* would be hired in his place. Entrepreneurs are similarly helpless to make decisions which might harm the machines, since the stockholders, the public and the government require efficient production from all manufacturing facilities for better dividends, lower commodity prices and more taxes. Any entrepreneur who made a decision to abolish the profitable machine (due to some notion about human survival, instead of profits) would be rapidly replaced by an entrepreneur who supported the development and use of profitable machines. Each man is, thus, individually helpless to change the course of events, and en masse he is equally incapable of changing the course of events. Man is helpless in the struggle against the machines.

MACHINES CONSUME MORE ENERGY THAN MAN. During the year 1970 the United States had a population of about 200 million people who consumed on the average about 2,400 kcal/day per person, yielding a total food energy consumption of about $7 \times 10^{24}$ ergs. During the same year utilities alone produced $55 \times 10^{24}$ ergs of electrical energy. But internal combustion engines and other devices consumed about seven times this amount of energy. Neglecting the energy expended by man for space heating, machines consumed about sixty times the energy for *food* as was consumed by man for his.

It might seem to be a matter of indifference as to what the relative consumption of energy was, except that man and machines are in competition. They play similar ecological roles. Their energies are directed in the same direction. If machines consumed sixty times the energy it means that their importance quantitatively in the ecosystem is sixty times greater than that of man. If machines have managed to become sixty times more important in the ecosystem in roles once dominated by man and his domestic animals in a brief 250 years, it presages the dominance of machines eventually in all roles played by man.

In terms of the ability to deliver mechanical power, man and his domesticated animals have been essentially extinct for at least the last fifty years.

THE BIOMASS OF MACHINES EXCEEDS THAT OF HUMANS. It is perhaps unnecessary to point out the obvious fact that the mass of machines far exceeds the mass of humans. But to stress the point, it may be noted that in the United States there is at least one automobile for every four persons (and in some areas there is more than one car per every two persons). Assuming that the average person weighs 60 kg or has 12 kg of carbon and the average automobile weighs 1500 kg or has a carbon equivalent of 250 kg, the biomass of automobiles is about five times the biomass of humans. But the biomass of automobiles probably represents only about 1 percent of the biomass of machines in general, so that the biomass of humans may be roughly estimated to be 2/1000 the biomass of machines in the United States. The much larger biomass of machines compared with humans is very significant because man and machines are competing. Mass-wise the machine has already won and man is already relatively extinct.

MACHINES ARE INVENTED TO REPLACE MAN. One of the most ironic features of the struggle for survival between man and machines is the fact that man invents and designs machines for the explicit purpose of replacing men. The value of a machine is often rated in terms of how many men it can replace, the more men replaced the *better*. An entrepreneur who converts from men to machines for a certain task with an attendant reduction in costs is frequently proud to state precisely the number of men the machines have replaced. The profit to the inventor and to the entrepreneur is, thus, directly proportional to the number of men destroyed as far as their ecological role is concerned. Since the human purpose of inventing and using machines is to replace men, it is not surprising that the final end result of such a process will be the extinction of man.

MACHINES WILL SHORTLY EVOLVE BRAINS SUPERIOR TO MAN'S. It has been generally felt that man has a secure position in the future scheme of things because he can out-think machines. But this will not always be true. Within the near future, perhaps less than fifty years, machines will be able to out-perform man in this area also.

In order to discuss thinking devices it is convenient to outline the thinking processes as follows:

A. Input
    1. Physical stimuli affecting the device
    2. Raw data input
    3. Immediate data reduction yielding perceptions

B. Central control
    1. Reduction of perceptions by pattern recognition yielding beliefs
    2. Choice of problems and choice of programs to solve them
    3. Mathematical and logical manipulations to yield solutions

C. Output
    1. Solutions reduced to concept of physical acts necessary
    2. Physical acts performed
    3. Physical effects produced such as a printout

Assuming the human eye has about $10^8$ rods and cones and that it takes a *picture* once every fifteenth of a second, the two eyes of a man can receive approximately $10^9$ bits of information per second. Considering the other sensory cells in the human body, a human being can receive perhaps $10^{10}$ bits of information per second. This capacity far exceeds the capacity of any computer or multichannel analyzer at the present. However, all of the bits of information that a human processes are not independent. They are largely wiped out as independent bits at the next level of analysis, namely the perceptual level. Human perception is rigidly built in, it cannot be changed; therefore, a human being can never realize the full informational potential of the bits of information received. A machine, on the other hand, can be designed with a flexible perception, so that the rate that a machine can receive bits of information is a better gauge of its actual capacity. Assuming that human perception operates such that $10^7$ bits of independent information survive the perceptual level, a machine that could handle $10^7$ bits of information per second should be able to match the human performance.

In the field of pattern recognition the human being is still the champion. There is, however, no cause for comfort in this fact, since it merely means that the inventors have not really spent much time on the problem as yet. (For some research in this area see the results of the Pisa Study [Grasselli, 1969].) It has already been reported in the literature that the visual field of a microscope can be divided up and digitized as to location and intensity of light received and this digital information programmed to yield the ratio of eosinophilic white blood corpuscles to the singly nucleated ones. The performance was superior to the performance of human laboratory technicians doing the same blood counts. There has also been some work done in computerizing cloud forms and correlating the results with weather. The present superiority of humans in the area of pattern recognition may be quickly wiped out when real economic advantages of computerized pattern recognition are exploited.

The process of abstract thought or problem solving requires a large fraction of the human nervous system, perhaps about one-fifth with about $10^{10}$ neurons. This may involve as many as $10^{13}$ synapses. If transistors and magnetic memory cores are to be equated to neurons or synapses, it does not appear that a practical computer could be constructed large enough to match a human brain. However, considering the fantastic development in technology over the last thirty-five years from the first vacuum tube computers to the present-day devices, there is reason to hope that the microminiturization of integrated circuits (e.g. Eimbinder, 1971) with appropriately automated production and assembly might yield a computer with an equivalent of $10^{10}$ elements in twenty years.

There is a strange aspect of the present matter that needs to be considered. If the human computer which resides in everyone's skull is so superior to electronic computers of even the most advanced design, why were mechanical desk calculators, vacuum tube computers, and solid state computers ever introduced in the first place? If man can out-think a machine why have all of the vast credit systems in the United States been computerized, Why was the author's bank statement in the 1930's and 1940's frequently wrong, while today it is never wrong? It would appear that man has already been replaced by superior thinking machines that know how to do arithmetic. It appears that no matter how huge the human computer might be, it is capable of delivering only vague outputs to trivial problems. The most elementary computer can multiply two 9-digit numbers and give the answer without error in less than a second. Why cannot a human?

The answer apparently lies in the evolution that gave rise to the human being as he is today. The large human brain was absolutely necessary for survival, anything smaller and man would have perished like his progenitors. Once this pertinent fact is clear, it also becomes clear that the human computer was evolved uniquely to solve the problem of man's survival. If the problem to be solved did not contribute to the survival of man it was rejected by a built-in program that rejected useless problems. There had to be the utmost economy of thought in order to survive, the brain being only as large as was absolutely necessary. There could be no time for frivolous problems. Thus, a *solution* for the human brain is always a solution to the single problem of survival, the problem of how to stalk a woolly mammoth, where to find the wiley antelope, which obstreperous neighbor to kill and how, on which side does one attack a cave bear, etc. The large number of neurons and synapses in the human brain are not necessarily available for true generalized thought, and, like the input branch, a factor of $10^{-3}$ times the number of neurons or synapses might better represent the number of neurons or synapses actually available for true abstract thought.

Since the human computer has evolved only to solve the problem of human survival hundreds of thousands of years ago, it is too rigid to satisfy the modern needs for flexible problem solving. The simplest and most stupid device with any degree of generality or flexibility such as the old mechanical desk calculator, easily outperforms man. In conclusion, it may be observed that, while the human brain is still probably the largest mathematical or logic device existent (neglecting whales) and it can outperform present-day machines in some ways, it is not sufficiently flexible to offer any competition for electronic computers once they increase about one to two orders of magnitude from their present size. In point of time this may occur within twenty years.

Finally there is the problem of how do the results given by a thinking device get translated into action. Once it has been ascertained what should be done, the precise train of events that are set into motion should follow a programmed sequence.

The output of abstract concepts per unit time, or printout rate is much greater for computers than for man. The latest photographic printout methods in bits per unit time probably exceeds the human output capacity by a factor of at least $10^6$, considering man's rate of typing on a mechanical typewriter or man's rate of speaking coherently into a recording device. Man's obvious inferiority in this area greatly reduces man's overall performance as a thinking device.

LIMITED SUPPLY OF FOSSIL FUELS FOR MACHINES. One hope that man might survive despite his insufficiencies compared with machines rests upon the idea that the limited supply of fossil fuels to feed machines will eventually become exhausted, leaving only carbon-based life with its ability to recycle carbon. Unfortunately, even if machines were restricted to feeding on recycled carbon compounds, their greater efficiency would permit them to replace carbon-based life. For example, during World War II in Europe it was found economically advantageous to use the charcoal taken from living trees to produce carbon monoxide to power internal combustion engines. By 1830 the efficiency of the steam engine exceeded that of all carbon-based prime movers, so that no matter how expensive the fuel, a heat engine would yield the greatest profit.

The potential direct utilization of solar energy with a possible increase in efficiency of perhaps ten times would also circumvent the need for carbon-based life. The direct utilization of nuclear energy, especially in breeder reactors and thermonuclear engines, could yield a flood of energy for machines. The limited supply of fossil fuels does not appear to place any restrictions on the eventual dominance of machines over man.

THE VERSATILITY OF A SINGLE MAN COMPARED WITH A SINGLE MACHINE. Perhaps men will be able to survive for many years yet because an individual man can perform a large variety of tasks. It might take a long time for a machine to evolve that has as small a mass as a man and can perform all of the tasks that can be performed by a single individual man. Unfortunately, machines do not have to evolve a single individual machine to perform all of the tasks that can be performed by a single human being. The principle of proliferation of species indicates that numerous machines playing complementary roles will be more efficient and will, therefore, be able to replace man.

LEARNING, MAN VERSUS MACHINE. It is sometimes felt that man's excellent ability to learn is unique to man and that it guarantees man a future over machines. But this is not true. In the first place, learning is of dubious value. If the end result that is learned is already known then a thinking machine should be programmed to handle problems without the wasteful time-consuming process of learning. In the second place, if the circumstances are so unique and peculiar that learning is, in fact, necessary, then computers can be programmed to learn in precisely the same way humans learn (Nilsson, 1965 and 1971). Evolution tends to short circuit the necessity for learning, since programming by experience involves a much larger computer capacity than is needed for a preprogrammed device. The learning process does not appear to confer upon the human species any particular superiority over machines.

WILL MACHINES IGNORE MAN? In the previous Chapter it was indicated that there are existent today species from all along the phylogenetic tree. It was argued that older species have evolved the best form to fit their primitive niches and therefore survive today. On this basis it might be felt that man might survive as a more primitive species than machines. But this cannot be the case. Machines have presented a sudden jump (a quantum jump) in efficiency and evolutionary rate. What is the *most efficient* for carbon-based life is still far less than the ideal efficiency potentially available for machine life. No carbon-based species is immune to replacement by a rationally conceived device, no matter what the niche. There is no niche in which man will continue to be better adapted. Eventually man will be out of a job.

The modern world that has unleashed nuclear energy, that has uncovered countless secrets of nature, that can circumvent the globe in ninety minutes, and that has put men on the moon still has Stone Age head hunters in New Guinea and the Amazon. By analogy, why can't machines simply overlook man in some places and not replace him entirely? Unfortunately, Stone Age savages are disappearing rapidly. The lateral diffusion of modern culture might seem slow, but it is not. The industrial revolution will come to New Guinea and the Amazon within the next few years. While man might go unnoticed by machines in out-of-the-way spots on the earth for a short time, he will be found and displaced eventually. Man has no place to hide.

It was indicated in Chapter V that outer space in orbit about the sun was an ideal habitat for machines. If this is true, then it might be reasoned that machines might go into space and leave the surface of the earth to man. But this cannot happen, since the general evolutionary principles will continue to operate on the surface of the earth as well as in outer space. The ecomass of the surface of the earth will attain a maximum with machines; it cannot be left to the less efficient human.

## Myths about Machines

The awareness of man's obvious inferiority to machines has been masked by a number of egocentric myths. Practically no one believes seriously that machines will make man extinct (e.g. Terborgh, 1966; Silberman, 1966), and when asked how they know that machines will never make man extinct, they are usually content to give one of the following brief statements:

MACHINES CREATE JOBS. This *non-sequitur* has a very comfortable sound to it. It serves to allay the fear of losing a job by being replaced by a machine, since somehow there is another job waiting that was created by the machine. A new factory in the neighborhood means more work for everyone. But, in fact, over any reasonably long period of time and over any reasonably large area it is discovered that the number of jobs is quite independent of the number of machines.

Since the amount of thermodynamic order that can still be produced on the earth's surface is virtually limitless, there will always be a job for every man as long as man exists. Over the years the number of jobs available will, consequently, just equal the number of men capable and willing to work. Similarly, if the word "job" is equated to "livelihood," it may be noted that in a population that does not starve the number of livelihoods just equals the number of men who have to feed themselves and dependents.

However, it should not be thought that merely because machines can never increase (or decrease) the total number of jobs available to man that machines do not replace man. Machines do replace man. Machines are taking over all of the significant thermodynamic functions in the ecosystem, leaving only trivial jobs for most men. Although a man's occupation may have only an insignificant thermodynamic effect on the ecosystem, he will still regard it as important to himself and to the world of men. Eventually, when there are not even any insignificant roles left for man, he will become replaced entirely.

MAN CONTROLS MACHINES. There does not appear to be anything particularly wrong with this statement at first glance; but the converse statement "machines control man" also appears to be true. A laborer working on an assembly line must fulfill the requirements of the machine, or else he runs a risk of losing his job and his capacity to feed himself. The machine, thus, controls the laborer. The entrepreneur is equally controlled by the machine, since it dictates the way the factory must be run. Since no individual man appears to have much option about how the machine is treated, it is difficult to see how man in general has control. There appears to be no point in the process of man's extinction by machines where any individual or men en masse can exercise an option to alter the process. However, it should be noted that there exists extensive literature which makes claims to the contrary (e.g. Ehrlich, 1972; Commoner, 1971).

MAN MAKES MACHINES. This is an error. Practically all of the effort, thermodynamically speaking, is provided by machines in order to make machines. Machines make machines. Presently man functions in the process primarily in decision making, a function that is being rapidly taken over by computers. There is no reason to believe that man will necessarily play a role in the making of machines in the future.

MAN INVENTS MACHINES. As has already been mentioned, the inventor's role will be increasingly taken over by computers until man will eventually be replaced. Even computers will be designed by computers (e.g. Herskowitz, 1968). If this statement, "man invents machines," is interpreted to mean that man, because he designs machines, will not permit anything bad to result from them, then it implies a peculiar motivation for inventors. Inventors, and therefore man, actually design machines to replace man, not to save man from the machines. The invention of war machines also indicates man's inability to control the situation from the inventor's side.

MAN NEEDS MACHINES. The symbiotic relationship between man and machines today is perfect. While man does not really need machines, since a population reduction by a factor of 100 (which could be achieved in one generation) would obviate the necessity of machines, it is still a *good* life that the machines provide. Each individual usually finds himself more or less necessarily tied to machines. Unfortunately, this situation cannot last. The symbiosis will disappear when man can no longer offer the machines any advantages. When the time comes man will appear to be more of a parasite than a symbiont to the machines, and the machines will then get rid of him.

MACHINES CANNOT BE CREATIVE. Some people retain the belief that man will survive because of his ability to create and invent which machines cannot do. Unfortunately, creative activity can also be programmed into a computer. Some computer-created music (for this theory see Mathews, 1969) is being played and appreciated. A mechanical device was invented almost two hundred years ago that could compose tunes. It was able to generate some five thousand tunes without repetitions. Modern computers can provide for no repetition at all. Computers can draw pictures (Siders *et al.*, 1966). Creativity is not a unique capacity of man. It is merely one facet of problem solving. When a problem requiring a creative solution is correctly analyzed, it can be programmed into a computer.

MACHINES CANNOT THINK. There are apparently no accomplishments of the human mind that cannot, in principle, be duplicated and improved upon by either present-day computers or those that will be developed in the near future. Machines *can* think (e.g. Feigenbaum and Feldman, 1963; Fink, 1966; Sayre and Crosson, 1963).

MAN MUST PROGRAM A COMPUTER. It is frequently reasoned that a computer cannot take over the role of a man because man is necessary to program the computer. The computer can do nothing more than a man, because man feeds in the program to be run. At the present it does appear that man plays an important role as a programmer, but this role is quickly disappearing. Once a complex program has been solved, it becomes software, a black box or tape which can be fed into any computer and no programmer is ever needed again for this particular program. As the pool of programs grow, and as computers themselves are used more often to generate more complex problems, the role of the human programmer vanishes. A programmer is like an inventor; his role can be duplicated by computers, and human programmers will eventually be replaced.

### Man's Immediate Future, a Prediction

Predictions for the near future should rely upon mathematical models, (e.g. Lave, 1966; Rashevsky, 1968). A number of studies that project current growth rates into the future, such as the Club of Rome Report (Meadows *et al.*, 1972), the M.I.T. Group (Wilson and Matthews, 1970), and the First International Conference on Environmental Future (Polunin, 1972), indicate some catastrophies within the next thirty years or so. These studies indicate the need for man to take action in order to survive or to preserve the quality of life. The basic difficulty, however, resides in the machine and not in man. It is the machine that pollutes the air, water and earth. It is the machine that is making so many species of carbon-based life extinct. It is the machines of war that threaten mankind. Thus, for man to take action he would have to control the machines; but this he cannot do. The machine is now the more dominant species; it cannot be controlled. Machines will not go away, and their biomass will continue to increase.

As long as machines require man and are not completely autonomous, the viability of the man-machine complex entails an adaptation by machines to guarantee the survival of humans. In the immediate future cars, factories and smelters will be equipped with more and more expensive devices to decrease pollution. The human population will decline to conserve resources and to avoid further pollution. However, the machine population will continue to grow. As machines gain greater and greater autonomy and the dwindling human population in the ecosphere decreases further and further in importance, machines will ignore human needs more and more. Pollution will eventually rise unchecked. By the time machines become almost completely autonomous humans will probably be restricted to buildings with controlled environments. After machines achieve complete autonomy, human extinction should follow fairly rapidly.

It is very hazardous to guess a time scale for these events, especially since they involve guessing about the rate of new inventions. There is a general tendency to assume certain devices or inventions are impossible only to find them available the next day. It is probably safe to assume, however, that machines will not be able to achieve complete autonomy in less than fifty years, and it would be foolish to think that it would take over two hundred years for machines to gain complete autonomy, considering the rapid exponential growth of machines over the last two hundred years. Thus, a rough guess would be about one hundred years to complete machine autonomy and perhaps two hundred years for human survival.

Needless to say some men will struggle for man's survival with every human device they can muster (e.g. Goldsmith *et al.*, 1972; Cellarius and Platt, 1972). The next few years may see some successes in their fight for a better environment and life for humans, but in the long run it appears hopeless.

# CHAPTER X

# TERRITORIALITY, A STRUGGLE FOR SURVIVAL

O NE FUNDAMENTAL NECESSITY of carbon-based life is sunlight. Since sunlight is provided at an average fixed rate per unit area, every existing organism (carbon-based) must command a certain portion of the products from a minimum area of the earth's surface or perish. A green plant must extend its leaves over at least a minimum area in order to obtain the sunlight necessary to live. A deer must eat brush off a sufficiently large area to provide its minimum caloric requirements for survival. The number of men that can live in a certain region is proportional to the area of arable land in the region. Each living organism is, consequently, vitally concerned with territory, the territory which will allow it to survive. The total area of an ecosystem (or the products from the total area) must become apportioned into individual territories that are large enough to permit individual survival. The question of territoriality is then the question of how land (or, equivalently, products) is apportioned to achieve survival.

### An Individual Must Fight to Survive

It is a general rule derived from field observation that no two species ever will appear to fill precisely the same niche. This could seldom result from competition between different species; instead it probably results from the simple fact that the requirements of the ecological niche forces all deviants of the same species to evolve toward the same end result, the most efficient structurally and behaviorally to play the role demanded by the niche. The true struggle for survival is, thus, not between different species, but between individual members of the same species. Each individual member of the same species requires precisely the same food and plays precisely the same role in the ecosystem. The real threat to the survival of a particular individual organism is primarily from its own kind. If an organism cannot keep members of its own species off of some minimum territory necessary for its survival it will perish.

This struggle between members of the same species (and less frequently against other species) takes on all of the aspects of warfare. Every individual organism is pitted against all other organisms (especially members of the same species) who are competing for the territory that is necessary for its own survival. This warfare is carried on with the most destructive weapons available to the individual; daggerlike eye teeth, crushing mandibles, poisons injected through stingers, poisons injected through hollow teeth, poisons sprayed into the air, crushing blows from massive appendages, sharp horns, pheromones to produce panic, sharp quills and spines, poisons injected through spines, poisoned flesh, pheromones to simulate a friend, poisons introduced into the soil, sharp beaks, electric discharges, release of obnoxious odors, and countless other devices and subtrafuges. Death is frequent (although perhaps only a small percent of aggressive encounters). The struggle to maintain a minimum territory, and therefore life itself, requires a vicious ejection or destruction of all competitors by whatever means available.

## How Aggressive Territorial Behavior Maximizes the Biomass

The result of the bitter deadly struggle for territory and survival is not, as might be supposed, chaotic. It performs some useful functions. It serves to eliminate defective genes. The healthiest and strongest inherit the earth; the weak and genetically deficient are deprived of their territories and perish. The struggle also serves to maximize the biomass of the species supported on a given area as required by Equations VIII.10 and its corollaries. Each healthy individual maintains just enough territory for its own survival and no more. Each individual organism owning only a minimum territory guarantees a maximum population for the species in the ecosystem.

The important question then arises as to how this can come about. It seems paradoxical that each individual, struggling for greater territory, results in less area for each.

POPULATION PRESSURE. An individual animal will try to maximize its energy consumption to maintain its mass and to optimize its internal thermodynamic order (*see* Chapter XII). In order to increase its energy consumption, an animal must increase the size of the territory upon which it lives. But to increase its territory, it must reduce the territory of its neighbors. The animal is therefore motivated to reduce its neighbors' territories. At the same time its neighbors are motivated to reduce its territory. The result is a *population pressure* (defined below in greater detail) which becomes established throughout the population. In equilibrium this population pressure is the same everywhere. If gradients in the population pressure arise, migrations of the species will result until all gradients again vanish. The uniformity of the population pressure throughout the total area guarantees that all individual territories will tend to be of the same size (assuming each area provides the same sustenance). As the population increases through natural reproduction, each individual area shrinks to a minimum. The net result is a maximum population for the total area.

To illustrate this phenomenon the behavior of passengers in a subway train may be considered. At a particular stop a large crowd of passengers rushes into a subway car and forms a tightly pressed mass near the door. As the train moves off and as time passes the passengers move away from the high population pressure region near the door (a pressure gradient existing) to establish a more-or-less uniform density and population pressure throughout the car. At the next stop the procedure is repeated. If the population pressure by the door after the first stop were not relieved, it would not be possible to add more passengers at the second stop. The behavior of humans to equalize the population pressure, thus, provides the capability for the car to carry the maximum number of passengers. Eventually, the car may get as tightly packed as possible to allow for survival, and no further increase in the number of passengers can occur, the maximum biomass having been attained.

"IDEAL GAS LAW" FOR POPULATION PRESSURE. In order to maintain a territory the boundary of the territory must be visited on occasion in order to be in a position to repell would-be trespassers by using pheromones, hostile displays or overt attacks. A quite general result may be derived by considering the special case of an animal that patrols its boundary. The ability of an owner of a territory to actually repulse trespassers at any point on the perimeter of the territory depends upon the frequency of visits by the owner to the point in question. If the frequency is too low a trespasser might cross the boundary, steal some forbidden fruit, and leave before the owner gets around to discovering the trespass. The time to patrol the perimeter of a territory once $t_p$ equals the length of the perimeter p divided by the mean velocity of the patrolling animal, v, or $t_p = p/v$. If an animal devotes a time $\tau_p$ per day to patrolling, then the frequency of visiting any point on the perimeter per day $f_p$ is given by $f_p = \tau_p/t_p$. Noting that the perimeter is proportional to the square root of the area of the territory A, $f_p \propto \tau_p v/\sqrt{A}$. The effect of patrolling the boundary upon the animal's neighbors should be a function of this frequency; the square of the frequency is chosen here as the measure of *population pressure*, thus,

$$P = \tau_p^2 v^2/A , \qquad (X.1)$$

where $P$ does not have the units of physical pressure (A constant could be introduced to provide proper physical units for P, but it would add nothing here). The similarity between Equation X.1 and the usual kinetic theory derivation of the ideal gas law (Sears, 1953) is apparent. This result, Equation X.1, may be regarded as being quite general for all techniques for maintaining territories, since $\tau_p$ may be taken as a measure of the daily effort to maintain the territory A no matter what the technique.

Perhaps the square of $f_p$ is not an adequate measure of the *population pressure*. Only observations can yield the best functional choice; however, the present choice appears to be sufficient for the present purposes. For this choice the pressure increases inversely as the size of the individual territory and increases directly with the total population on a fixed area. It increases as the square of the patrolling activity as measured by $\tau_p v$. These relationships appear reasonable for any concept of pressure.

### Symbolic Aggression and the Need for Actual Physical Conflict

It has been noted by a number of observers, especially Lorenz (1966 and 1952) and Tinbergen (1966), that many species have evolved complex organs and behavioral patterns for ritualized displays which replace actual physical conflicts. Such ritualized displays, while preserving the struggle for territory, apparently minimize the number of destructive conflicts and allow the healthiest, i.e. the winner, to escape unharmed. Ardrey (1966) has written with great literary skill about various examples of such territorial behavior.

A much earlier work by Allee (1938) is of interest because of the many experimental results he presents which demonstrate that animals in groups tend to fare better than singly. The world view of Allee appears to be incompatible with the concept of territorial aggression, yet many of the experiments that he reports may have resulted from aggression rather than cooperation, since territorial aggression can also lead to beneficial results (as already noted above). For example, a Siamese fighting fish does not prosper and may, in fact, die, unless it can see a hated rival. Aggression and conflict, as it ordinarily occurs in nature, should not be viewed as necessarily detrimental. On the contrary, aggression and conflict is undoubtedly a valuable asset to a species. Such an unpeaceful view is perhaps hard for a man to accept who is accustomed to thinking that he shuns conflict in order to resolve his own personal problems. Such a *reasonable* man may be inclined to believe that nature must also resolve its problems without conflict. Nature has not, however chosen peaceful solutions; conflict and aggression are apparently the most efficient mechanisms for apportioning a population over a fixed area such as to maximize the biomass.

It might be concluded after reading such authors as Lorenz, Tinbergen, Ardrey and Allee that symbolic aggression has evolved to replace actual physical conflict. This cannot, however, be entirely the case; actual physical conflict must always remain and must always play a significant role. The most that territorial advertising through displays of bright feathers, fins, songs, pheromones, sharp teeth, hackles, etc., can achieve is merely a reduction in the rate of physical conflict. Physical conflict can never be replaced entirely. This conclusion may be made clear by considering a hypothetical example. A bird with genetically deviant characteristics presents an outlandish display of bright feathers, but also has a weakened musculature. This hypothetical example is then a case of false advertising; the threatening bird is incapable of backing up its threats in physical combat. It is evident that in a very short time any such outlandish display would come to mean a weak insufficient individual who could be easily attacked and physically ejected from a territory. It may, thus, be assumed that physical conflict always remains to insure the fact that the advertising display corresponds to the actual physical ability of the advertiser. The success of ritualistic behavior and aggressive display in deciding territorial questions, thus, depends upon the accuracy of the information imparted. The biggest, gaudiest bird had better be the best fighter in an actual physical conflict. Field observations have generally supported the fact that individuals with the greatest display are also the most sufficient individuals aggressively and physically.

One of the reasons that the significance of actual physical combat has been underestimated is that very few aggressive confrontations ever result in an actual physical combat. Almost all aggressive confrontations are resolved by ritualistic displays. It has, thus, been frequently assumed that evolutionary development replaced physical conflict by ritualistic displays. But no matter how successful the ritualistic display may be in apportioning territory, eventually true physical conflict must occur to test the competing organisms. The small number of physical conflicts that result from aggressive confrontations should not be used as a measure of the importance of actual physical conflict. No matter how infrequent, physical conflict is important. For example, neighboring countries may engage in threats and negotiate for fifty years and then one day an actual physical war breaks out with a duration of perhaps only a few months. The results on the humans involved is more likely to be measured in terms of the unlikely war of a few months duration than in terms of the frequent symbolic aggressions over the period of fifty years. Even if only one aggressive confrontation in a thousand or even ten thousand results in physical conflict, the rate of physical conflict may still be significant.

## Some Observations of Physical Conflict in Nature

Despite the author's limited experience as an observer of nature, he has become impressed with the importance of physical conflict in nature. During three years of maintaining a bird feeder outside of his dining room window where the birds could be observed at a distance of no more than two meters at all hours during the day, it was found that no individual bird (all birds being individually identifiable) failed at some time or another to physically attack another bird of the same species. Activity at the feeder was very great and all birds showed a constant fear and concern about being physically assaulted. About thirty-five different species of birds regularly visited the feeder, and there were close to four hundred individual landings per day in the feeder. The species of bird that appeared to be the most guilty of direct physical assault was the linit, males being more aggressive than females.

In Mesa, Arizona, while convalescing from an accident, the author had the time and opportunity to observe the complex network of ant trails of three different species of ants. At one point the trails of two different species were observed to cross. From the author's previous observations he became convinced that it was a general rule of ant behavior that the ant trails of two different nests never cross, even if the trails were produced by two different species. The territorial rights of an ant nest to an adequate hunting and feeding territory is apparently preserved by this rule. By observing the anomalous situation, he found the larger red ants (about 1 centimeter long) would repeatedly attack and chase the smaller black ants (about 0.5 centimeters long) when they crossed the red ant trail. Both species were highly upset and perturbed; the hostility and friction at the intersection was very great. One day after about five days the author discovered a large army of red ants composed of many hundreds of individuals attacking the black ant's nest. The battle raged for at least 2½ hours (the limit of the author's endurance). The next day there was a pile of black ant corpses covering the ground around the nest for about 10 centimeters and about 1 centimeter thick. The pile must have contained many hundreds of individual black ant corpses. There were no survivors; even after observing the nest for five more days no black ant was observed to approach or leave the nest. The obnoxious and forbidden crossing of trails had been eliminated. The red ants continued to use their trail without interference. The red ant attack was not a *slave* raid, since the number of attackers was much too large and the defenders showed none of the usual panic. In addition, the species of red ant is the  most common in the Southwest and does not take slaves, and slavers do not kill the adult population of the nest they raid. The attack did not arise from a desire to use the black ants as food, since their corpses were left behind uneaten. The attack was apparently to preserve the territorial integrity of the red ant domain.

The author has observed many isolated individual combats between members of the same species. The most impressive instance (apart from human conflict) of apparently senseless slaughter was the case of the common California red ant on a vacant lot in Altadena, California. Over an area of about one thousand square meters, this one species of ant could be observed in mortal combat. Heads, legs, antenna, abdomen, pieces of the carapace, etc., could be seen as large red drifts in some sections of the area. During five different visits over a period of five years the scene always appeared precisely the same. The death rate must have exceeded several hundred individuals a day. It was impossible to discover the home nest of either of the competing factions; they seemed to each have nests randomly distributed over the whole area. Ten years after the first obser-

vation the author again revisited the area. This time no ant living or dead could be found after searching for two hours over
the whole area.

## A *Mathematical Model of Territorial Behavior*

It is of interest to enquire into how territorial behavior affects the life of an animal. How much of its life is devoted to
territorial defense? How often will it engage in conflict to protect its boundary? Under what conditions might it fail to keep
its territory and consequently lose the means to survive? To
try to find answers to such questions it is useful to first construct
some sort of idealized model. Such a model can yield insight
into what variables determine the animals' behavior.

The fundamental and obvious variable is *energy*, as has been
stressed frequently in this book (e.g. Chapters XI and XII). A
consideration of the energy involved in any process will generally
yield extremely important information about the nature of the
process. The second most important variable here is the *time*
(which is not independent of the energy expenditures). Thus,
a model for the behavior of a single territorial animal should at
least involve the time spent doing different tasks and the energy
expended doing these tasks.

In addition, any model should be based upon the principle
that any group of animals will behave such as to maximize the
biomass of the biosphere (Equations VIII.10 and its corollaries).

It will be assumed that steady state equilibrium has been
reached in the ecosystem and that the population pressure, Equation X.1, is a constant.

ENERGY BALANCE FOR A DAY'S ACTIVITY. Besides patrolling
the boundary of its territory, the animal must take time to hunt
for the food found in its territory; he must take time for grooming; he must take time on occasion for mating and caring for the
young; and he must take time to sleep. If $\tau_h$ is the mean time per
day spent in an active state apart from patrolling, then $\tau_h R_a$,
where $R_a$ is the mean active metabolic rate, is the energy burnt
up per day in all activities apart from patrolling. If $\tau_s$ is the time
spent per day sleeping or in a quiescent state and R is the rest-

ing or quiescent metabolic rate (close to the basal metabolic rate), then $\tau_s R$ is the energy dissipated each day in an inactive state. The energy assimilated each day E equals the number of kills per day N, or the mass of food eaten per day, times the energy assimilated per kill e, or per unit mass of food eaten. Conserving energy then yields

$$E = Ne = (\tau_p + \tau_h)R_a + \tau_s R , \qquad (X.2)$$

where the active metabolic rate $R_a$ has been assumed to be appropriate for patrolling, and $\tau_p$ is the time spent per day patrolling.

FOOD GAINED FROM HUNTING. The rate that food is made available per unit area on the territory is more-or-less fixed for most species (e.g. detritus feeders and grazing animals) as has been discussed in Chapter VII. The rarer and less important case of a species that feeds on live reproducing prey and thereby influences the rate at which food becomes available will not be treated here. If $\rho$ is the area density of food and if the rate that organisms other than the animal of interest consume this food per unit area is taken as proportional to $\rho$, then the rate that food increases per unit area becomes

$$d\rho/dt = P - b\rho - N/A , \qquad (X.3)$$

where P is the constant rate of supply per unit area, b is a proportionality constant, A is the area of the territory, and time rates are taken per day. Assuming that a steady state becomes quickly established, the right side of Equation X.3 is zero; or

$$\rho = \rho_0 - N/Ab , \qquad (X.4)$$

where $\rho_0 = P/b$ is the area density of food in the absence of the animal of interest.

An animal hunting for food distributed randomly over an area as discrete units has a certain lateral cross section $\sigma$ for finding and consuming the food, The amount of food eaten n in time t equals the number of food units lying in an area of width $\sigma$ and length vt where v is the speed of the hunting animal; or

$$n = \sigma \rho vt , \qquad (X.5)$$

(*see* Fig. 31). Thus, the number eaten per day N is given by

$$N = \sigma \rho \, v \, \tau_h, \qquad (X.6)$$

where the total active time apart from patrolling is assumed to be devoted to hunting, a sufficiently adequate approximation here considering the arbitrariness of the separation between active and quiescent times and corresponding metabolic rates $R_a$ and $R$. Similarly, it is sufficient here to assume that the same mean speed is used in both patrolling and hunting for food.

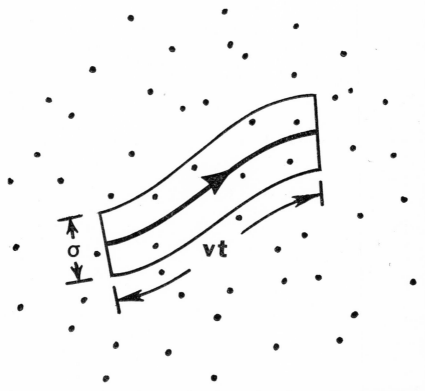

Figure 31. Diagram indicating that the number of prey taken in the time t are those in the area σvt.

Since the entire day is taken up with patrolling, being in an active state apart from patrolling, and sleeping,

$$\tau_p + \tau_h + \tau_s = 1. \qquad (X.7)$$

The active metabolic rate $R_a$ may be set equal to the quiescent rate R plus a term proportional to the velocity v,

$$R_a = R + av , \qquad (X.8)$$

where a is a constant of proportionality. For strenuous running this expression, Equation X.8, is not correct, but for the average daily activity it should be an adequate first approximation.

BEHAVIOR PREDICTED BY MAXIMIZING THE BIOMASS. A set of six equations (X.1, X2, X4, X.6, X.7 and X.8) has been accumulated which expresses certain necessary average features of the physical life of the territorial animal. These six equations contain eight constants, $P$, $R$, $\tau_s$ e, $\rho_o$ b, $\sigma$ and a, and seven variables, $\tau_p$, v, A, $\tau_h$, $R_a$, N and $\rho$. One more condition is required to determine the system. This may be obtained by requiring the total biomass on the total area to be a maximum. Since the mass of each individual animal is assumed to be constant in the present example (i.e. R is chosen to be a constant), the total population on the total area should be a maximum, or the individual area held by each animal should be a minimum; thus,

$$A \quad \text{is to be a minimum,} \qquad (X.9)$$

which yields seven equations in seven unknowns.

Substituting the value of $\tau_s$ from Equation X.7 into X.2, then substituting in the value of $\tau_p$ as given by Equation X.1 and the value of $\tau_h$ as given by Equation X.6, and finally using Equations X.4 and X.8 yields the result

$$a\sqrt{P\,A} + abA\,\eta/\sigma(1 - \eta) + R = b\,\rho_o A\,\eta\,e ,$$
$$(X.10)$$

where

$$\eta = 1 - \rho/\rho_o . \qquad (X.11)$$

This result, Equation X.10, is a relation between the two variables A and $\eta$, all other variables having been eliminated. Solving Equation X.10 for $\eta$ then yields $\eta$ as a function of A; thus,

$$2\,\eta = f + g - [(f + g)^2 - 4f]^{1/2} , \qquad (X.12)$$

where f is a function of A and g is a constant that are given by

$$f = R/\rho_o ebA + \sqrt{P}\,a/\rho_o eb\sqrt{A} ,$$
$$g = 1 - a/\rho_o e\,\sigma . \qquad (X.13)$$

To estimate the magnitudes of f and g, it may be noted from the equations presented above that

$$f = (1 - \rho/\rho_o)(1 - \varepsilon_h/E),$$
$$g = 1 - (\rho/\rho_o)(\varepsilon_h/E),\qquad (X.14)$$

where $\varepsilon_h = \tau_h(R_a - R)$ is the daily energy expended while active, apart from patrolling, in excess of the quiescent amount that would be expended during this time and E is the total energy assimilated per day. It may, thus, be concluded from Equations X.14 that

$$\eta < 1 \quad \text{and} \quad 0 < f < g < 1. \qquad (X.15)$$

To find the condition for a minimum area, Equation X.12 may be differentiated with respect to A (yielding the reciprocal of the derivative of A with respect to $\eta$) and the result allowed to go to infinity; thus,

$$2d\,\eta/dA = \{1 - [(f + g)^2 - 4f]^{-\frac{1}{2}}(f + g - 2)\}\,df/dA. \qquad (X.16)$$

Since df/dA never goes to infinity for finite values of A, according to the first of Equations X.13, the condition for minimum area is for the quantity in the square brackets to vanish or for

$$f = [1 - (1 - g)^{1/2}]^2. \qquad (X.17)$$

A Partial Numerical Example. There appears to be no satisfactory approximation that can be made to simplify the problem; therefore, it will be more illuminating to consider an explicit partial numerical example. From the second of Equations X.14 one possible value of g which appears reasonable is

$$g = 0.8. \qquad (X.18)$$

Introducing a dimensionless area,

$$\alpha = A/A_o \quad \text{where} \quad A_o = R/\rho_o eb, \qquad (X.19)$$

the expression for f from the first of Equations X.14 becomes

$$f = \alpha^{-1} + C\,\alpha^{-\frac{1}{2}}, \qquad (X.20)$$

where C is a constant. One possible value of C which appears reasonable is

$$C = \sqrt{\bar{P}} \, a / \rho_o eb \sqrt{A_o} = 1 \, . \qquad (X.21)$$

The value of $\alpha$ for which the area is a minimum, as given by Equations X.17, X.18, X.20 and X.21, is found to be 4.07. From this result and the equations above the following numerical values may be obtained:

$$\alpha_{min} = 4.07 \, , \quad \eta = 0.553 \, , \quad \rho/\rho_o = 0.447 \, , \quad f = 0.306 \, ,$$
$$\varepsilon_h/E = 0.477 \, , \quad \varepsilon_p/E = 0.0786 \, , \quad R/E = 0.444 \, , \qquad (X.22)$$

where $\varepsilon_p = \tau_p(R_a - R)$ is the energy expended while patrolling less the quiescent amount that would have been expended during the same period.

If in addition to the two numerical assumptions above, Equations X.18 and X.21, it is assumed that the animal in question sleeps one fourth of the twenty-four-hour day, then the following numerical values may be obtained:

$$\tau_s = 0.250 \, , \quad \tau_h = 0.644 \, , \quad \tau_p = 0.106 \, , \quad R_a/R = 1.668 \, ,$$
$$av/R = 0.668 \, , \quad E_p/E = (\varepsilon_p + \tau_p R)/E = 0.126 \, ,$$
$$E_h/E = (\varepsilon_h + \tau_h R)/E = 0.863 \, , \quad E_s/E = \tau_s R/E = 0.111 \, ,$$
$$(X.23)$$

where $E_p$ is the energy expended per day while patrolling, $E_h$ is the energy expended per day while active apart from patrolling, and $E_s$ is the energy expended per day while sleeping. All of these numerical values, Equations X.22 and X.23, appear to be reasonable possibilities for a predator such as primitive man. Only three of the original eight constants have been arbitrarily assigned here (g, C and $\tau_s$), leaving five constants (R, b, $\sigma$, e and $P$ to still be specified to completely determine the problem numerically.

It may be noted from Equation X.23 that for the numerical values chosen the animal expends 13 percent of its total assimilated energy in patrolling its territorial boundary. This might appear at first to be excessive, yet many individual species appear to expend even a larger fraction of their total energy in the territorial defense effort. For example, *Homo sapiens* in the United States over the last twenty years have undoubtedly exceeded this percentage. Including not only the effort to preserve the national territory but also the effort to preserve internal territories and property (property, being generally derivable from territory, may generally be equated to territory), the percentage of the gross national product over the last twenty years devoted to the national military (at least 13%), national investigatory agencies, state police, county sheriffs, city police, private guards, watchmen, locksmiths, prosecutors, judges, bailiffs, and all of the other agencies and personnel associated with essentially patrolling functions has undoubtedly exceeded 20 percent.

In conclusion, from simple energy and time requirements and the assumption that territorial animals behave such as to maximize their biomass, a description of how the territorial animal spends its day is obtained. The partial numerical example here yields numerical values for the fraction of the day spent hunting (or active apart from patrolling) $\tau_h$, sleeping $\tau_s$, and patrolling $\tau_p$. The fraction of the total energy expended in these three activities is also given. If the remaining five constants were specified, all of the remaining variables would be determined, such as the actual size of the territory held A, the velocity while patrolling v, and the amount of food eaten in a day N. A further numerical example (modified for a social species) is presented below. More realistic and detailed considerations of the various activities considered here would yield more realistic results, but the theoretical procedure would remain the same.

## A Mathematical Model of a Territory Held by Primitive Man

It has been pointed out by Lorenz (1966) that man is an aggressive territorial animal. Even preman, such as Australopithicus, living in fixed sites and hunting game, must have had to struggle with his neighbors to maintain a sufficiently large territory upon which to survive. Primitive man, who may still be observed in some parts of the world today, lived, and still lives, an active aggressive territorial life. Primitive man formed social units varying in sizes from just an immediate family to villages of perhaps several hundred individuals.

A mathematical model for human territorial behavior may be obtained by slightly modifying the discussion at the beginning of this Chapter which concerned the behavior of a solitary territorial species. In particular, letting n′ be the number of individuals sharing one territory, Equations X.1, X.2, X.4, X6, X.7 and X.8 become modified to read:

$$n'\,\tau_p v/\sqrt{A} = \sqrt{P}\,, \qquad (\tau_p + \tau_h)R_a + \tau_s R = Ne\,,$$
$$\rho = \rho_o - Nn'/bA\,,$$
$$N = \sigma\,\rho\,v\,\tau_h\,, \qquad \tau_p + \tau_h + \tau_s = 1\,, \qquad R_a = R + av\,,$$

$$(X.24)$$

where N is now the number of kills per individual human, the total kills equalling n′N, and where $\tau_p$, $\tau_h$ and $\tau_s$ are the fraction of a day spent by the population on a single territory patrolling, hunting (i.e. active) and sleeping, respectively. The discussion above still applies if b is now replaced by b/n′ and P is replaced by $P/n'^2$. Assuming that the partial numerical example considered above, Equations X.18, X.21, X.22 and X.23, is also a valid example in the present instance, and assuming that it is possible to estimate the remaining constants for the case of primitive territorial man, a complete numerical example may now be explored.

The present example envisions a very small village of primitive humans that subsist by hunting deer. It is assumed that the boundary of the hunting territory is maintained by patrolling. It is not necessary to assume that patrolling has to be frequent, since a recent trespass is readily discernible to a trained hunting eye and retributions can be exacted upon an offending neighbor by attacking the neighbor's village. If it is assumed that the deer are of such a size that one kill will keep twenty humans alive for one day, then $N = 1/20$. Since it may be roughly assumed that it takes 2500 kilocalories per day to keep an active human functioning, $Ne = 2500$ kcal/day. From these two assumptions $e = 50{,}000$ kcal/kill. If the mean velocity while hunting or patrolling, as averaged over the whole day, is taken to be about 1 mile/hr, which is much less than a steady walking pace, then $v = 50$ km/day. On the average, primitive man probably caught and killed a deer within a fairly short lateral range, kills at a distance of 100 meters probably being extremely rare. A mean effective lateral cross section for killing deer of about 30 feet, or $\sigma = 0.01$ km, is probably not unreasonable.

The value of the constant b may be estimated from the following considerations: If a doe has one fawn per year, the reproduction rate per deer is 0.5 per deer per year. This means that $1.37 \times 10^{-3}$ deer are added per day to the population of deer per deer. Since b is the fraction of deer dying per day due to all causes other than human predation, man eats deer at a rate of $(1.37 \times 10^{-3} - b)$ deer per day per deer. From $Nn' = (1.37 \times 10^{-3} - b)\ \rho\ A$ and the third of Equations X. 24, using the value of $\rho/\rho_0 = 0.447$ from Equations X.22, the value of $b = 1.37 \times 10^{-3}\ \rho/\rho_0 = 6.12 \times 10^{-4}$ deaths per day per deer. This result yields about $\rho\ A/n' = N(1.37 \times 10^{-3} - b) = 66$ deer per human and implies that $Nn'/1.37 \times 10^{-3}\ \rho\ A = 0.553$, or about 55 percent of all deer born are eaten by humans and about 45 percent die from other causes.

In order for the results to be reasonable for the present numerical example, it is necessary to assume that the population of the village is very small; in particular, the value of n' is chosen as 8. Summarizing the numerical values that have been assumed:

$$N = 1/20 \text{ kills per day per person}, \quad \sigma = 0.01 \text{ km},$$
$$v = 50 \text{ km/day},$$

$$Ne = 2500 \text{ kcal/day person}, \quad b = 6.12 \times 10^{-4}$$
$$\text{deaths/day deer}, \quad n' = 8.$$

$$(X.25)$$

Using these numerical values, Equations X.25, X.18, X.21, X.22 and X.23, the following numerical quantities may also be derived:

$$e = 50{,}000 \text{ kcal/kill}, \quad \rho = 0.155 \text{ deer/km}^2,$$
$$\rho_0 = 0.347 \text{ deer/km}^2,$$
$$R = 1670 \text{ kcal/day}, \quad a = 22.3 \text{ kcal/km}^2,$$
$$A = 3370 \text{ km}^2,$$
$$R_a = 2780 \text{ kcal/day}, \quad \sqrt{\bar{P}} = 0.731 \text{ per day}, \quad (X.26)$$
$$r = 32.7 \text{ km},$$

where r is the radius of a circle with the area A, and where the value of $\sqrt{\bar{P}}$ represents, for a circular territory, a frequency of about once every five days at any point on the boundary of the territory. Primitive man actually hunted all kinds of game, and an important part of his diet was of vegetable origin, so that actual territories were undoubtedly smaller and supported more individuals than would be indicated by this numerical example. Yet this example does serve to indicate that primitive man had to sustain himself upon a large territory in order to survive. Primitive man was necessarily a rare species per unit area compared with herbivores. Unlike modern man, it might be speculated that primitive man's territorial hostility rarely led to deadly conflict due to his low population density and the infrequency of confrontations. However, over sufficiently long times the population pressures apparently selected modern man over the Neanderthal (even if modern man is a descendant of the Neanderthal), since the Neanderthal became extinct.

## Man's Territorial Instincts

Species that aggressively maintain a territory for a single individual, or at most for a family group, can easily identify strangers or individuals that should be attacked and driven from the survival territory. *Homo sapien,* being both a social and a territorial species, has a problem: How can members of the in-group or home village be distinguished as members of the *we,* and members of all other groups and villages be distinguished as *they?* Primitive man, like all other social species, evolved patterns of behavior to help, succor and to love the *we* and to attack, expel and hate the *they.* Practically all social species other than man, such as ants, bees, rats and wolves, use pheromones to distinguish between the *we* and the *they.* An animal lacking the correct nest scent is generally attacked and may be killed unless it manages to escape. For example, removing a rat from its home colony, changing its scent, and reintroducing it into its colony results in its death by its nest mates tearing it to pieces; they believe it to be one of the hated *they.* Humans must distinguish between the *we* and *they* primarily by sight. Besides slight family resemblances that man can easily recognize, primitive men use special articles of clothing and decoration and tattoos to aid in the identification.

It is interesting to note that not only does man face a society without, he also faces a society within. A man's body constitutes a huge society of many billions of individual differentiated cells which must also be able to recognize the difference between *we* and *they*. All foreign tissue that is introduced into the body without being digested is recognized and is ultimately rejected and repulsed by the immunological response. This effective response is calculated to keep out all would-be intruders such as bacteria, virus and larger parasites. It is also calculated to limit growth and sustenance to the *we* as opposed to the *they*. It may be conjectured that one of the reasons that the entire chromosomal content is passed on to each and every differentiated somatic cell is in order to help make the identification of *we* and *they* certain. Only between identical twins with the same genes and chromosomes (assuming no mosaic divergences) is there no immunological response. Both twins are tagged with the same identification for *we*.

Direct observations of primitive villages verifies the hostility of one village toward its neighboring villages. Neighbors are generally hated, loathed and feared. Neighbors are not to be trusted; they are deceitful; they are not at all the honest straightforward type as those in the home village. Very primitive villages are comprised almost solely of closely related individuals. Frequently the differentiation between *we* and *they* then rests upon the presence of a blood relationship. Great care is taken to keep the traditionally held hunting territory intact against any encroachment by poaching neighbors. Hostile displays, verbal threats, agreements, ruses, open conflict and cannibalism all play a part in maintaining and adjusting territorial boundaries.

The sexual act in humans appears to be much too frequent to concern reproduction alone. In most species reproduction is achieved with efficiency and economy of effort. In most mammals pheromones indicate the readiness of the female to be impregnated, and these pheromones then stimulate the male to perform the requisite act. Thus, a brief period with a few copulations is generally sufficient for each pregnancy. On this basis, an efficiently reproducing human male should copulate with any one female no more than a few times a year. Moreover, it appears to be the height of inefficiency for copulation to continue even after a female has long since conceived and even for decades after the female has passed the menopause and is no longer fertile. As has been indicated previously, species evolve toward higher ecological efficiencies, which implies reproduction with a minimum of effort. It may therefore be concluded that man's frequent copulation must serve another purpose besides reproduction. In most species the strongest bond established between two individuals of the same species is cemented, fulfilled or indicated by ritualistic *grooming* behavior. The specific rituals performed and the organs used and stimulated vary greatly among different species. Considering the pair bonding that frequently occurs between copulating human partners, it may be postulated that the sexual act in man is primarily a *grooming* act rather than an act performed solely for procreation. The act establishes a *love* relationship between a man and a woman (and less frequently between two men or two women). The bond between such grooming human partners can last for decades. The significant point for bringing up the present discussion is that the grooming partner is always a member of the home village and is always accorded a recognition of *we*. The bond established by *grooming* with only one partner then establishes a bond with the entire home village. The pleasure and gratification of the grooming act then tends to make an individual feel that all that is best and good is possessed by *we* and not by *they*.

Under certain circumstances it may be imperative for primitive man to distinguish as rapidly as possible between *we* and *they*. Away from the home village and near the territorial boundary one of the *they* might attack, while one of the *we* might give aid. It could sometimes be a matter of destroy or be destroyed, involving a choice within a split second. As a consequence, there tend to be no subtle differentiations between *we* and *they*. Everyone confronted needs to be immediately classifiable. No matter how big the brain might be in the human skull, it has been evolutionarily programmed to yield answers concerning survival only. In crisis situations the only answers are elementary: run, freeze, kill, feign death, attack, present hostile display and retreat, etc. Modern man's huge brain does not, therefore, exempt him from the same instinctual behavior patterns of his ancestors and other territorial animals.

Modern man has, thus, inherited the strong instinct to dichotomously classify all persons as either *we* or *they*, *good guys* or *bad guys*, *friend* or *foe*, a member of the *ingroup* or *outgroup*, *countrymen* or *aliens*, *comrades* or *strangers*, *friends* or *enemies*, etc. In times of war most of us have personally felt the hatred for the *enemy* and the love for one's own country. Such emotions do not arise from any abstract intellectual effort; they arise from our territorially derived human and prehuman instincts. (Inversely, accepting such emotions without attempting to override them with a rational intellectual effort may have survival value.)

The fact that men do from time to time kill each other in disputes indicates that killing fellow members of *Homo sapiens* is within the acceptable repertoire of man's instincts. From direct observations it is obvious that man is a killer, not only of game, but also of his fellow man. Australopithicus, a human precursor, and primitive man have left among their bone deposits an impressive storehouse of information concerning their violence and cannibalism. It may be speculated that any lone hunter who happened to wander far inside his neighbor's territory was probably hunted down, killed and eaten with relish by those whose territory he had dared to violate. It may be noted that cannibalism, which is apparently an intrinsic part of human nature, has a number of advantages: It puts some food into the stomach, it limits the population, and it helps to keep the neighbors at an appropriate distance.

Sadism, torture, cruelty, and the most exotic and painful meththods of destruction of fellow members of *Homo sapiens* are also part of the repertoire of ordinary human instincts. Any normal man or woman when properly stimulated and encouraged by the social approval of his fellows from the same ingroup will derive extreme pleasure from watching or participating in the torture and degrading destruction of a despised, captured and helpless enemy. The direct observation of primitive man in colonial North America and in the present-day Amazon and New

Guinea, as well as the information presented by sophisticated cultures such as the games of ancient Rome, the concentration camps of Hitler Germany, and atrocities in South Vietnam and in Bangladesh (formerly East Pakistan), amply document the sadistic nature of man.

The occasional occurrence of the lone sadist also supports the idea that sadism is part of the instinctive nature of man. A lone individual who, without social approval, experiences pleasure and gratification by torturing, destroying and/or eating members of his own ingroup is considered to be a monster who is generally imprisoned or killed with great dispatch when discovered. Such a person is sick, not because he has the correct impulses to kill, maim and torture, but because he does not wait for the social approval usually necessary and he chooses as victims members of his own ingroup.

Since instinctive patterns of behavior arise by evolutionary selection, the cruelty and torture by humans must have some survival value. Perhaps the degrading destruction of a despised and helpless enemy heightens the hatred of the enemy and increases the desire to kill the enemy who are still free and who still represent a threat. Sadism may, thus, serve to heighten the resolve and dedication to destroy the hated enemy. Such sadism may also serve to increase the emotional distinction between *we* and *they*.

In the comfort of the home village where each individual generally has some solicitude for his fellow man, it is hard to believe that man could have such sadistic instincts. It may be readily appreciated that humans would be unable to maximize their population per unit area if torture and cannibalism were very frequent events in terms of ordinary man hours or in terms of the fraction of victims per unit time compared with the replacement rate of the population by natural reproduction. But the rarity of sadistic acts and cannibalism does not imply that man lacks the instincts for cruelty or that such rare acts are not important. Similarly, the rarity of physical conflict and war in comparison to the frequency of hostile confrontations does not imply that man lacks a killing instinct or that physical conflict and war are not important. Man's nature is neither pure love nor pure hate; it is both: He loves his family, home, friends and country and hates strangers, foreigners, foreign countries and enemies.

Perhaps one of the most remarkable bits of evidence of man's aggressive territorial instincts is provided by his unusual hair distribution. Animals that leave pheromones to mark boundaries seldom have need for showy visual displays to advertise their ownership. The raised hackles on a dog, the arched back of a cat, and the hiss of the opossum are modest displays compared with the bright plumage and songs of birds that must rely upon sight and sound to advertise their ownership of a territory. Humans having poor noses compared with other mammals must also rely upon sight and sound to distinguish between *we* and *they*. It might therefore be expected that man would use sight and sound to advertise his ownership of a territory. The human male has evolved a beautiful display of long bushy head hair and a beard. The beard is a flag to warn off trespassers. Children and women do not have beards for the very good reason that they are not generally involved in maintaining territorial boundaries. The human male's beard is a complete analogue of the male lion's mane, the male cardinal's brilliant red topknot, the male baboon's ample ruff, the male sea elephant's proboscis, and all of the other displays put forward by male animals in species where the male aggressively maintains the territory. Modern man, thus, carries on his person the flagrant display of his aggressive territorial nature. The peculiar fact that the beard hairs that grow on the neck grow upward while the face hairs grow downward is now explained: the neck hairs make the face hairs stand out in a showy display. The bald head that occurs in some healthy human males (and not females) does not necessarily represent an evolutionary reduction in the display of hair; on the contrary, since the amount of hair per person tends to remain more or less constant, baldness probably implies a redistribution of the available hair to yield a larger, showier beard. As mentioned previously, evolutionary selection guarantees that such territorial advertising does not lie; thus, within a genetically homogeneous group of human males those that can grow the biggest, showiest heads of hair and beards should be the largest, healthiest, strongest and most dominant. As in the displays of male birds, it is extremely unlikely that the beard is a sexual sign to attract a female or even a sign to help distinguish males from females.

Callan (1970) presents a brief but interesting review of the ethology of human aggression.

## Frequency of Conflict in a Territorial Species

Although the fraction of hostile confrontations which result in actual physical conflict may be ordinarily extremely small, there must be, on the average, a proportionality between the frequency of physical conflicts and the frequency of hostile confrontations. The rate of confrontations and, therefore, the rate of conflicts which occurs in a territorial species may then be estimated by noting that the frequency of confrontations should be approximately proportional to the rate that two neighbors chance to meet within a certain *confrontation distance l* on their common patrolled boundary. In particular, a given length $l$ on the perimeter of the territory is occupied by a patrolling animal for a part of each day equal to $t_p l$ p, where $t_p$ is the time per day devoted to patrolling the boundary and p is the length of the perimeter of the territory. The chances that a neighbor, possessing the same size territory, will also be on this distance $l$ of the common boundary at any time of the day is equal to this same expression. The chance of en encounter per day on the length $l$ is then given by the product of the chance that the animal will be on the length $l$ times the chance that his neighbor will also be on the length $l$, or

$$(t_p l/p)^2 . \tag{X.27}$$

The chances of an encounter are the same for each length 1 of the perimeter of the territory held by a single animal, so that a single animal will experience confrontations at an average rate of p/l times the value given by Equation X.27, or

$$t^2_p l/p . \tag{X.28}$$

But from Equation X.1 the frequency of visiting a point on the perimeter is a constant, so that the rate of confrontations is proportional to

$$(Pl/v^2)p . \tag{X.29}$$

Assuming $P$, $l$ and $v$ are constants, the conflict rate per day per animal Q becomes proportional to the perimeter of the territory,

$$Q = \beta p , \qquad (X.30)$$

where $\beta$ is the proportionality constant.

The total rate of conflict $Q'$ on an area $A'$ will equal Q times the number of individual territories held (divided by two in order not to count each conflict twice); thus,

$$Q' = pA'/2A = \beta P , \qquad (X.31)$$

where A is the area of an individual territory and P is the total perimeter around all individual territories (taken only once) on the area $A'$. If there is a natural external boundary to the area $A'$, such as the shoreline of an island, the outside perimeter should not be included.

The frequency of conflicts per unit area varies inversely as the perimeter around an individual territory; thus,

$$Q'/A' = \beta p/2A = \beta'/p , \qquad (X.32)$$

where $\beta'$ is a constant of proportionality. Since the area density of population d is simple $1/A$, it may be concluded that the frequency of conflict per unit area is proportional to the square root of the area density of the population; thus,

$$Q'/A' = \beta''\sqrt{d} , \qquad (X.33)$$

where $\beta''$ is a constant of proportionality. This confirms the usual observations that the frequency of conflicts per unit area increases when the population density increases.

While the mathematical arguments in the present section have been based upon a species that patrols its boundary in order to maintain its territory, the results are probably applicable to all aggressive techniques for maintaining a territory. Whatever is the mechanism for keeping neighbors from crossing boundaries of a territory it works both ways, so that the frequency of encounters to maintain a territory must necessarily be proportional to the perimeter of the territory. Another way to appreciate this conclusion is to consider an animal surrounded by neighbors occupying adjacent areas of one fixed size. If the animal in question enlarges its own territory it will gain a number of new neighbors directly proportional to the increase in the perimeter of the enlarged territory. Assuming that the confrontation rate is generally constant with each neighbor, then the confrontation rate is increased directly as the number of new neighbors or as the length of the perimeter of the territory. No matter what the particular devices are for maintaining a territory, the frequency of confrontations and, therefore, conflicts is then generally proportional to the perimeter.

## Frequency of Wars and Territoriality*

According to Equation X.31, it was concluded that the total rate of conflict should be proportional to the total length of the perimeter around all contested territories. In particular, territorial behavior indicates that conflict should be almost exclusively limited to neighbors. Assuming that wars result from territorial behavior, the frequency of wars between countries should then correlate with the number of neighbors a country has. Lewis Fry Richardson (1960a) showed that this is, indeed, the case. He found that the number of external wars between 1820 to 1945 with more than 7,000 war dead correlated with the number of frontiers for the thirty-three countries he investigated.

This correlation, while demonstrating that the effect of territoriality exists, does not indicate the precise magnitude of the effect. To evaluate the situation more accurately it is possible to proceed much as Richardson (1960b) did. It may be noted that wars of a given size will usually be fought where the population of the smaller side sustains a loss of at most some fraction k of its population. Thus, the smallest population that can generally be expected to engage in a war with a total of n war dead is

---

* The mathematical results of this section were first published in a paper entitled "Frequency of Wars and Geographical Opportunity" by J. P. Wesley in the *Journal of Conflict Resolution*, 6 (4):387-389, December 1962. Permission to reproduce this material has been granted by the Publisher, Sage Publications, Inc. The paper was subsequently reprinted in *Theory and Research on the Causes of War* (Pruitt and Snyder, 1969). These results are presented here yet again because the originally proposed explanation in terms of *geographical opportunity* is now recognized as being too trivial, the true explanation appearing to lie in man's territorial behavior.

n/2k, it being assumed that both sides suffer about the same number of casualties, n/2. If the population of the world is broken up into cells whose populations are each of this minimum size, then there will be at most s potential belligerents that might engage in a war with n war dead where

$$s = 2kW/n \,, \qquad (X.34)$$

where W is the world population.

In terms of territoriality it may be assumed that only neighboring cells will go to war against each other. Richardson (1960c) compared the number of common boundaries or frontiers between neighboring cells with the frequency of wars of different sizes but failed to obtain precise agreement with observation. The number of boundaries is not, however, a proper measure of the frequency of hostile confrontations, for if two countries share a long common boundary they will have a greater frequency of confrontations than if they share a short common boundary. As already derived, a proper measure of the frequency of confrontations is the total length of frontiers or boundaries between all of the population cells. The measure chosen here is in population units and does not involve actual physical length. A long physical frontier between two countries with low population densities might afford the same rate of confrontations as a short frontier between two countries with high population densities.

If A is the total land area of the earth, then each cell may be assumed to occupy an area $a = A/s$. The perimeter about each cell is proportional to $\sqrt{a}$. Summing over all of the s cells then gives a total perimeter about all cells which is proportional to $\sqrt{s}$,

$$P \, \alpha \sqrt{s} \,. \qquad (X.35)$$

From Equations X.34 and X.35 the total perimeter P about all s cells is seen to be proportional to $n^{-1/2}$,

$$P \, \alpha \, n^{-1/2} \,. \qquad (X.36)$$

It is now postulated that the number of war dead generated per unit time is proportional to the rate of confrontations as measured by P, Equation X.36. If df/dn is the frequency of wars producing war dead in the range from n to n + dn, then the rate at which war dead are produced in wars of this size is given by

$$n \, df/dn \propto n^{-1/2} . \tag{X.37}$$

Equating this rate of generation of war dead, Equation X.37, to the total confrontation rate, Equation X.36, it is found that

$$n \, df/dn \, \mu \, n^{-1/2} . \tag{X.38}$$

In terms of logarithms Equation X.38 may also be written in the form

$$\log_{10}(df/d\log_{10}n) = C - 0.5 \log_{10}n , \tag{X.39}$$

where C is some constant. This relation is precisely the same as the empirical relation already established by Richardson (1960d) whose summarized data for wars between 1820 to 1945 are reproduced in Table VIII. The constant C has been chosen equal

TABLE VIII. FREQUENCY AND MAGNITUDE OF WARS

| War dead $n$ | Magnitude $\log_{10}n$ | Number of wars $df/d\log_{10}n$ | Observation $\log_{10}(df/d\log_{10}n)$ | Theory $C - 0.5 \log_{10}n$ |
|---|---|---|---|---|
| $10^3$ | 3 | 198 | 2.30 | 2.34 |
| $10^4$ | 4 | 70 | 1.84 | 1.84 |
| $10^5$ | 5 | 24 | 1.38 | 1.34 |
| $10^6$ | 6 | 6 | 0.78 | 0.84 |
| $10^7$ | 7 | 2 | 0.30 | 0.34 |

to 3.84 so that the theoretical curve coincides with observations for wars involving $5 \times 10^3$ to $5 \times 10^4$ war dead.

It may be noted that, while the important result Equation X.36 was derived using rather abstract arguments, it can, in principle, also be checked directly from maps where the area of each country is made proportional to its population. The total perimeter about countries in a given population range should then vary as the negative one-half power of the population.

## Population Control and Territorial Behavior

It has been proposed by some observers, especially Wynne-Edwards (1962 and 1965), that territorial behavior has evolved primarily for the purpose of population control. It is argued that a species must exhibit population self-control, or otherwise its population would increase unchecked, it would eventually destroy its own food supply, and it would then subsequently become extinct. It is claimed, for example, that lemmings and other rodents by aggressively maintaining individual territories larger than they apparently need for their immediate survival do not threaten to destroy their own food supply by overgrazing and thereby guarantee the survival of the species. During the process of raising their young many birds establish nesting territories. The entire area upon which the species lives becomes divided up into a fixed number of nesting territories. It is claimed that this number is fixed by the aggressive instincts of the species involved and that each nesting territory is more than adequate to maintain the nesting pair and its young. It is assumed that the number of young produced per year per nest remains constant. It is, thus, argued that the number of young raised each year over the entire area becomes fixed independent of the population, thereby yielding a stable population control mechanism as indicated in Chapter VII. Similarly, certain species, such as sea elephants, some gulls and seals, breed only in very limited fixed areas which can apparently be apportioned only among a certain fixed number of breeding females, thereby yielding a fixed number of offspring per year. Other species, such as the sage grouse (Scott, 1942), apparently limit their reproduction rate by limiting copulation to certain arenas with a limited number of males at only certain times of the year.

While this plausible picture may have some elements of truth to it, it does not appear to be basically correct. Most species cannot eat themselves out of house and home because they cannot influence the rate that food is made available to them. Most species (e.g. insects) do not show any particular population self-control and can exhibit rapid population increases of many fold with subsequent crashes. Many devices for population control exist in nature which do not imply any self-control. Most herbivores have populations significantly controlled by predators. Some species utilize the negative feedback provided by cannibalism. Density-dependent variables such as disease may limit the population of other species. Thus, population self-control cannot be a general feature of all species, whereas, territorial behavior to maintain the minimum area necessary for survival is present in all species.

In order to maximize the time average biomass, a species population should tend toward a maximum. This implies that the territory held by each individual should be as small as possible for survival. If a species were, in fact, to threaten to destroy its own food supply, then it is conceivable that the territories held might become larger than the bare minimum for survival. However, this would probably be an unusual case involving a rare species. It may be speculated that the reported restraint by certain species in maintaining a population less than the maximum may merely reflect the observer's imperfect knowledge of all of the necessary features needed for survival. For example, lemmings and other rodents may have to maintain a certain territory for *runs* to avoid predators which exceeds the minimum feeding territory.

The tendency for a species to reproduce at a maximum rate to take advantage of time varying environmental conditions (such as mosquitoes reproducing in a pond in the spring after the temperature has attained the appropriate level) appears to be counter to the concept of population self-control. The crowding together of breeding sea elephants, gulls and seals may not represent any self-restraint on the reproduction rate; it may impart protection to the breeding females or it might merely reflect some peculiarly favorable conditions for breeding. Such species may be actually optimizing their reproduction rates.

The concept that nesting territories are larger than necessary for the breeding pair and offspring is probably also in error. For a maximum biomass the nesting territories should be just the bare minimum possible. The purpose of aggressive apportionment of nesting territories should be to make each nesting territory no larger than is just necessary for raising a brood. In this way a fruitful area can support more individual territories and yield more offspring than an unfruitful area of the same size. The clutch size and the number of broods per year do not remain fixed but instead vary depending upon the particular yearly conditions (Lack, 1969). Observations tend to support the more reasonable hy-

*Ecophysics*

pothesis that birds raise as many young per year as they are
physically able.

The struggle for individual territories tends to force a maximum utilization of an area. Under constant environmental conditions an area will then tend to support a constant population and
will, thus, tend to yield a constant number of offspring yer year.
The territorial struggle, thus, results fortuitously in a population
control mechanism. This fortuitous consequence of territorial behavior does not imply, however, that territorial behavior evolved
as a result of a need for population control.

# CHAPTER XI

# A MEASURE OF PSYCHOLOGICAL BEHAVIOR

THE STUDY OF ECOPHYSICS AND ECOLOGY involves the study of how living things interact with each other and the environment. Therefore, the natural behavior of organisms in their natural habitat is of central interest in ecophysics. The prediction or explanation of the natural behavior of a single organism also becomes important. It might appear that the study of the behavior of an animal is *psychology* and not *ecophysics*. Strictly speaking, this is correct, but only certain particular aspects of psychology are of special interest in ecophysics and not the whole field of psychology. An animal is to be viewed as a *black box* in ecophysics, a *black box* that exchanges matter and energy with the environment and may move to alter the environment and interact with other *black boxes*. Ecophysics is involved with the external aspects of behavior only.

The internal structure of sensors, nerves, ganglia, brains, etc., is of no interest here. There is no interest here in the details of the internal processes that are necessary to permit an organism to do what it does. The learning process, being generally a transitory phenomenon in the ontological development of an organism to fit it into its niche, is similarly of little interest here, since the final behavior is dictated by the organisms' ecological role, whether learning took place or whether the organism was preprogrammed with the appropriate instincts. Laboratory experiments that do not involve a duplication of a natural habitat or niche are of no interest. In ecophysics laboratory observations must be validated against field observations, since the behavior of animals in a laboratory can be subject to many factors which can be easily overlooked and which can seriously affect the results.

The psychology of interest to the ecophysicist might appear to be superficial to a psychologist, since it involves little of the detailed information generally of interest to psychologists. Nevertheless, the psychology of interest to ecophysicists is fundamental in the sense that all natural behavior (and the mechanisms that give rise to this behavior) is dictated by the ecological role of the organism. The motivations that an organism has must be perfectly matched to its needs as required by its niche. The psychology of motivation, perhaps the most basic area in psychology, therefore, comes within the purview of ecophysics.

The present Chapter and the following Chapter discuss psychology in terms of the behavior of an organism in its natural habitat subject to its natural needs. The emphasis is on the all-important energy and time conditions that the behavior of an organism must satisfy in its niche (e.g. the energy and time requirements used to predict territorial behavior as presented in Chapter X). The application of physical theory remains a central theme of this development. Perhaps the term *ecophysical psychology* as applied to this research would not be entirely inappropriate.

## The Need for an Externalized Concept of Behavior

It is frequently said that psychology is the study of behavior. The word "behavior," however, denotes and connotes different things to different individuals. Even though it may be agreed that behavior is what an organism does, it is not always clear just exactly what is meant. For example, it may be agreed that a man who sits motionlessly in a chair with his eyes wide open is behaving, but there may not be universal agreement as to what he is doing. It is frequently felt that a statement about the man's external appearance and activity is insufficient to prescribe what he is doing. Instead the man's behavior, or what he is doing, is frequently specified by such statements as: He is watching a movie, he is watching television, he is worrying about bills, or he is waiting up for his teenage daughter to come home.

Although it may be convenient in everyday life to classify and label behavior according to such projected or ascribed motives, purposes, goals, emotions, perceptions, knowledge, etc., it is not the appropriate rigorously scientific thing to do. Scientifically it is essential to differentiate between what is observed directly and what is inferred indirectly. Although it may be easy to avoid *anthropomorphisms* when studying inanimate systems or lower organisms, it becomes increasingly more difficult when dealing with more advanced animals or human beings.

Here a rigorously scientific viewpoint will be adopted. An organism will be regarded as no more nor less than the physical system that it is. The word *behavior* will be chosen to mean only the direct external observation of the physical activity of an organism, and the word *behavior* will not be used for inferred internal states. For example, the *behavior* of a man who is sitting motionlessly in a chair with eyes wide open is specified simply as the activity of sitting motionlessly in a chair with eyes wide open. If a television set is in line with a normal line projected from the cornea of the man's eyes, it might be hazarded that the *explanation* of his behavior is that he is *seeing* or is *watching* the television screen. Since he could not be watching television unless he were sitting motionlessly (ordinarily) with eyes wide open, in some sense the *explanation* of the *behavior* also specifies the *behavior* itself. Despite the fact that the economy of everyday language lumps together such different facets of our knowledge and experience, it is necessary here to distinguish sharply between the gross physical aspects of an organism and its activities and the inferred causes or explanations of such activities.

### Behavior as a Dynamic Concept

The word *behavior* implies a specification of an activity or a time (either implicit or explicit) varying function. If a man is sitting motionlessly or is sleeping this is *behavior* since the bodily processes are dynamic and such quiescent states are of only a finite duration.

In contrast to the dynamic implications of the word *behavior*, there are also properties of the system that may be easily observed that do not imply activity such as mass, volume, height and temperature. The entire prescription of the state of an organism must include both the dynamic properties or *behavior* and the static properties or characteristics.

The simplest dynamic states are *steady states*. A *steady state* may be defined as a state of a system that involves the time implicitly and not explicitly. For example, a river may be characterized by its width, depth and by its volume of flow; consequently, a river is in a *steady state*. The flow of water is a dynamic process, yet, once the volume of flow has been specified, the time does not have to be considered explicitly. The conception of *behavior* usually implies such a steady state of indefinite duration. For example, when one envisions a dog *running*, it does not require an explicit consideration of the time.

Even though it may not be possible to characterize all actual behavior as a steady state, the simplicity achieved by ignoring explicit reference to the time frequently makes it useful to consider *ideal behavior*—*ideal behavior* being a strictly steady state phenomenon. Thus, even though a steady state is not adequate to describe all activities of interest, it frequently yields a very valuable first approximation.

### Change of Behavior

If a running dog stops and starts scratching, the dog has *changed* its behavior. The dog has changed its behavior from the steady state of running to the steady state of scratching. It is frequently felt that psychology involves primarily a prediction of such *changes* in behavior. This view implies that little or no attention should be given to the study of steady states per se, but instead it implies that the primary attention should be directed toward the explanations and predictions of changes from one steady state to another steady state. There can be no question that some of the most interesting predictions possible will involve the prediction of such changes of behavior. However, the only way to adequately predict such changes in behavior is to have an understanding of the steady states themselves. Without considering a change in behavior it may be asked what supports, causes or maintains an indefinite continuation of a particular steady state.

## Periodic Behavior

Steady state behavior is a very important concept, but actual animals do not maintain a steady state for any great period of time. Animals behave, in general, in a recurrent fashion. Some behavior is repeated periodically in a very regular manner. For example, almost all animals have numerous acts, such as sleeping, that are repeated with the very definite period of twenty-four hours. The twenty-four-hour period for the amount of sunlight and for temperature variations imposes a twenty-four-hour period (a circadian rhythm) upon most recurrent behavior.

In addition to the very definite periods, such as the twenty-four-hour period, there are other recurrent activities that have a more random quality. For example, a horse might urinate five times a day, but the precise times during the day when the horse urinates might not be predictable. This type of behavior might be classified as randomly periodic or quasiperiodic.

Breathing and the heart beating are examples of recurrent behavior with very short periods. These periods are usually quite regular, but may be subject to aperiodic modifications as well as to temporary increases or decreases in the period.

In a sense it is also possible to treat periodic behavior as steady state behavior, because the time need not be explicitly included in the description of what an animal does once the period is specified.

Experimentally it is usually assumed that an animal has been kept on a regular daily routine which is then regarded as largely immaterial to the observations being made. Here, on the other hand, the routine or the complete twenty-four-hour behavior is regarded as highly significant. A study of the routine behavior is considered to be of greater importance than the study of new behavior that is illicited when an animal is required to respond to a new or peculiar laboratory situation.

## Choice of Continuous Rather than Dichotomous Measures of Behavior

When attempting a scientific description or explanation of a phenomenon, one of the first problems that must be faced is: What shall be observed? What observables are significant and what observables are not significant? Does a significant observable depend upon the nature of the measures used? If this is true, what is the nature of the measures that are significant? Although there have been studies classifying some of the different types of measures that may be used and that are sometimes fruitful, most such studies fail to properly indicate the types of measures that are the best.

Some observables have only dichotomous measures: For example, the sex of humans (neglecting freaks) is dichotomized as either male or female. Some observables have finite integer measures: For example, the number of legs animals have are integer and finite; mammals have four legs; insects, six legs; spiders, eight legs; etc. Some observables have integer but infinite measures: For example, there are a finite integer number of atoms in a microgram of substance, but an infinite number in an infinite universe. Some observables have only probabilistic measures: For example, the appearance of heads when flipping a coin has an expectation of one-half. There are many probabilistic measures that might be used to distinguish many different probabilistic observables. Some observables have continuous measures: For example, all magnitudes of length, time and mass can be observed in principle.

When attempting to choose the most valuable observable and its measure, there is one rule that transcends all others in importance. The use of this rule usually allows the choice to become obvious. The observables and consequent measures that should be chosen should be those that permit generalizations or scientific inductions from the fewest laboratory observations to the largest number of all possible observations. If an observation applies to one member of a dichotomous class, then the generalization to the opposite member is either trivially valid, is trivially false or is impossible. Whatever happens to be the case, dichotomous measures clearly present the weakest measures that can be chosen (unless one wishes to consider probabilistic dichotomous measures that are weaker still). An observation using a dichotomous measure permits a generalization to only *one* other case. The strongest measures that can be chosen are continuous ones, since no matter how small an interval is examined by a finite number of laboratory observations, generalizations can be made to fit an infinite number of possible cases. (The number of possible cases is, in fact, not actually infinite, since the error of measurement in effect reduces the number of different magnitudes that can be distinguished to just a finite—although possibly large—number.)

In terms of information theory the observations inside the laboratory and the observations outside the laboratory may be each analyzed, in principle, into so many *bits* of information. Assuming ignorance about the significance of each of the bits of information, the importance of each bit may be regarded as being equivalent to the importance of any other bit. The measure to be chosen should permit the largest number of bits of information to be predicted outside the laboratory for the least number of bits of information that have to be collected inside the laboratory. A dichotomous measure contains only two bits of information, while a continuous measure contains an infinite number of bits of information (in principle, assuming no observational errors). Consequently, a prediction of a continuous variable gives, in principle, infinitely more information than the prediction of a dichotomous variable.

On this basis the most significant observables to describe an animal's behavior are the continuous ones. For example, in addition to the continuous static observables such as mass and temperature, the animal's velocity while running, its rate of doing work, and its rate of energy dissipation must all be important observables. Whether an animal is drinking or not drinking, running or not running, defecating or not defecating, etc., being dichotomous measures are not of any great significance. It may, of course, be necessary to know the value of such dichotomous observables, but the predictive value or the basic understanding of phenomena provided by such dichotomous variables is necessarily meager.

To stress this important point still further the phenomenon of boiling water may be considered. The boiling or not boiling of water may be described as a dichotomous variable. To predict this observable one might wish to use dichotomous observables: The kettle must be placed or not placed on the stove. The fire must be lit or not lit under the kettle. It is necessary to wait a long time or a short time. All of the necessary precursors to boiling water may, thus, be catalogued. This catalogue would provide prediction. (As a matter of fact, this is precisely the form of prediction used in the kitchen to predict the boiling of water.) This approach does not, however, yield any insight into the actual causes or mechanisms of boiling water. The understanding of the infinite variety of conditions under which water will boil only arises when one turns to the continuous observables at hand. The conditions under which water will boil are completely specified by just two continuous observables—the pressure and temperature. (While aware of the continuous increase in the warmth of the water before it boils, it is not regarded as a significant predictive variable in one's own kitchen.)

To the primitive mind only the act of boiling and the acts preparatory to boiling are of interest, and without a knowledge of the pertinent continuous variables no understanding of the actual causes of boiling can ever result. Similarly, if dichotomous behavior is to be predicted only in terms of dichotomous observables, then little real understanding of the phenomena involved can ever result.

The present discussion may also be structured in terms of the desirability of having a quantitative science as opposed to a descriptive science. Since all qualitative or descriptive measures may be regarded as dichotomous variables, the desirability of a quantitative science as opposed to a descriptive science merely means the desirability of having continuous as opposed to dichotomous observables or measures.

A continuous observable can be placed into correspondence with a continuous number scale. With such quantitative measures it is then possible to employ the extremely fruitful mathematics that has been developed over the centuries. Since valuable results invariably follow from the study and measurement of continuous variables, the following rule may be proposed: *If there are continuous observables associated with a phenomenon, then it is extremely important that these observables should be measured.* No matter how irrelevant or trivial such continuous variables might appear to be initially, in the long run the continuous variables will be discovered to be the most significant variables.

### Intensity or Rate of Behavior

Besides a description of the activity which distinguishes one form of behavior from another—such as drinking from running—behavior also involves an *intensity, degree, amount* or *rate* of behavior. For example, not only can a dog run, it can also run at different speeds. Not only can a man lay bricks, he can also lay a greater or less number of bricks per day. All dynamic phenomena, which includes behavior, involve observables that have a variable time rate (i.e. generalized velocities). Since time is a continuous quantity, such variable time rates or velocities, such as the rate of flow of water in a river, will have (in general) continuously varying magnitudes.

Since the intensity or rate of behavior is a continuously varying observable, one of the most significant and important things to measure, according to the previous subsection, is this intensity or rate of behavior.

### Intensity of Behavior as the Time Rate of Expending Energy

A measure of the intensity of behavior should be applicable to all activities or behavior, in order to have the greatest generality. Only by having the same measure of the intensity of behavior for all activities is it possible to compare the significance of different activities. It is only with such a common measure that complete scientific significance can be given to the idea of *intensity of behavior.* Most measures of the rate of behavior are clearly inappropriate. For example, the velocity of running cannot be chosen as a proper measure of the rate of picking coffee berries, or the speed of climbing a mountain cannot be used to measure the rate of typing. Continuous variables such as the temperature, volume, mass and pressure are all measurable for all organisms and for all activities, but these particular variables measure static properties only, while the *intensity of behavior* requires a time rate or dynamic measure.

The time rate that an organism expends energy is found to have all of the properties that should be associated with the concept *intensity of behavior*. This observable satisfies the basic and necessary requirement that it can be readily measured in the laboratory with the proper equipment. It is a continuous observable. It can be measured for all possible activities, since, no matter what an organism is doing, all life processes involve the expenditure of energy. The time rate that an organism expends energy is a dynamic property.

Other observables can also fulfill the criteria of being measurable, being continuous, being measurable for all activities, and being dynamic properties. For example, the rate of breathing, the rate that carbon dioxide is generated, or the rate that the heart beats would appear to be equally satisfactory measures of the intensity of behavior. However, the time rate of expenditure of energy provides for the greatest possible degree of generality; it does not depend upon detailed knowledge of either chemistry or structure. The time rate of expending energy is a measure applicable to all species of animals and plants as well as all species of inorganic life. The choice of the time rate of expending energy for the intensity of behavior permits the greatest degree of generalization possible from laboratory experiments.

### Intensity of Behavior and the Rate of Doing Work

To illustrate the significance of the rate of expenditure of energy as the measure of the intensity of behavior, consider an organism that does external work. For example, consider a rat running in a squirrel cage such that it raises a weight to perform external mechanical work. Since most easily observed activities involve such a performance of mechanical work, most easily observed behavior has a rate of doing mechanical work associated with it. For many ordinary purposes the time rate of doing mechanical work would be a sufficiently good measure of the intensity of behavior. However, the rate of doing mechanical work, which may be more directly associated with what might ordinarily be regarded as the intensity of behavior, is compatible with the measure already chosen here, since the amount of work done by an animal will be proportional to the energy expended.

An animal will necessarily expend more energy than it does work, since no animal or thermodynamic system can be 100 percent efficient in converting chemical energy to mechanical work. Moreover, the functional relationship between the energy expended and the work accomplished is not linear; the rate of expending energy, or the intensity of behavior, increases more rapidly than the rate of doing mechanical work. It, thus, takes more than twice the rate of energy expenditure to double the time rate of doing external mechanical work. The time rate of expending energy, which can be directly determined experimentally, is a more fundamental measure of the intensity or degree of behavior than merely the rate of doing mechanical work, since many activities do not even involve the performance of external work.

## Thermodynamics and Behavior

There is still another way that it may be seen that the time rate of energy expended is the best and most appropriate measure of the intensity of behavior. From the viewpoint of physics all real objects in the universe may be regarded as physical systems. Therefore, an object that is an organism is also a physical system. Fundamentally when each of the atoms of a system is specified as to type and when each of the positions, velocities and mutual attractions are given, then the entire physical system is specified. Such detailed knowledge is rarely available, however, so that other measures are used to characterize the system. These measures involve summing or averaging (or both) over the effects of all of the atoms. There arise sum concepts such as the total mass or volume, and there arise average concepts such as pressure and temperature. Such statistical measures characterize the system as a whole, or in a macroscopically heterogeneous system such measures provide a description for each part of the system.

The two most fundamental and general laws that govern such measures of statistical systems are the first and second laws of

thermodynamics. Living organisms, while very complex heterogeneous systems, are still statistical physical-chemical systems
that must obey these first and second laws of thermodynamics.
Because energy is a concept that is fundamental to both of these
laws, any energy exchange that an animal engages in with its environment is of the utmost importance. The most basic and important scientific knowledge that can be gained about an animal's
activities must necessarily include a knowledge of such energy
exchanges. The identification of the time rate of energy expenditure with the intensity of behavior is, consequently, not entirely
arbitrary. This identification provides the concept of *intensity of
behavior* with the greatest possible generality and scientific
potency.

# CHAPTER XII

# BEHAVIOR PREDICTED BY
# ENERGY NEEDS

FROM THE FACT that the earth is evolving towards states of lower entropy or higher thermodynamic order, it was concluded in Chapter VIII that ecosystems evolve toward a maximum biomass, Equation VIII.11; and finally it was concluded, Equation VIII.13, that

*Each individual organism tends to create and maintain an internal state of maximum thermodynamic order.*

This conclusion is true on the average considering the behavior of organisms in general. It is not always rigorously true for any particular organism at any particular time or place. However, most of the failures to follow this general principle occur in *abnormal* situations, situations for which the organism is not genetically equipped. Therefore, the discussion in the present Chapter is limited to individual behavior as it occurs in nature under ordinary circumstances. It is assumed that an organism has been evolutionarily selected to play its role in its natural habitat, and that situations beyond the natural habitat or beyond the requirement of its role are outside of the scope of the present research. While certain specially selected laboratory experiments can be used to verify (or reject) theoretical models (such as will be suggested below) for predicting behavior in the natural habitat, laboratory experiments in general merely demonstrate the inability of an organism to respond appropriately to most artificial situations.

### *Two General Types of Behavior*

A description of what a particular animal is doing indicates the type of behavior involved. An animal is drinking, it is eating, it is running, it is standing, etc. Drinking, eating, running and standing are all examples of types of behavior. Behavior is classified here according to type depending upon the direct observable physical nature of the acts involved; presumed internal states will not be used. That the animal is slaking its thirst, satisfying its hunger, avoiding danger, or thinking are not appropriate descriptions of types of behavior, since they presume a knowledge of the internal state of the animal.

The basic behavior of organisms to create and maintain order in the ecosphere, Equation VIII.14, may be conveniently categorized into the following two general types (which may not always be strictly exclusive):

1. behavior which tends to maximize the thermodynamic order external to the organism and

2. behavior which tends to maximize the thermodynamic order internal to the organism.

It might at first appear that heterotrophs (and especially higher trophic level predators) behave such as to tend to reduce, rather than increase, the amount of order external to themselves. For example, foxes eat rabbits, which apparently is behaving such as to reduce the total biomass by apparently reducing the number of rabbits. But this is not the case, since rabbits reproduce to make up the deficit. Steady state life processes on the earth are cyclic, so that the mere eating of one organism by another does not necessarily decrease the time average biomass. Predation in nature generally permits a larger standing biomass. Similarly, behavior of Type 1 above does not preclude the possibility of competition of one deviant with another leading to the extinction of one of them, since the survivor will tend to have a greater ecological efficiency (i.e. a greater ability to support more biomass per unit energy consumed). Behavior such as to maximize external thermodynamic order also implies behavior to tend to maximize the reproduction rate to adapt to changing environmental conditions (*see* Chapter VIII). Behavior of Type 1 includes the possibility of altruistic behavior. A warrior ant (or human), for example, may unhesitatingly sacrifice its life in order to allow the entire nest the opportunity to maintain its population. A mother bear may defend her cubs unto death, thereby permitting the population of bears to remain from one generation to the next.

Behavior of Type 2 above is more important, since the thermodynamic order created and maintained is ordinarily internal order. The behavior of all organisms, including humans, is primarily concerned, therefore, with maintaining internal order. The remainder of this Chapter will be concerned only with behavior that tends to maximize internal thermodynamic order.

### Maximization of Energy Expended to Create and Maintain Internal Order

The mass of ordered compounds in an organism is a measure of the amount of order being maintained internally (roughly the dry mass). To create and maintain an internal state of maximum thermodynamic order then means a maximization of the internal mass of the organism. The constraints on the organism are such that a certain size cannot be exceeded without interfering with other necessary functions of the organism. The maximization of the mass must be done while taking into account simultaneously all such constraints. This means that the mass should be regarded as a dependent variable which is a function of everything happening to the organism. Mass should not be regarded as simply a direct observable which is easily measured and is used merely to render all quantities mass specific.

The amount of mass m that can be maintained internally, i.e. the amount of thermodynamic order, is in general proportional to the amount of energy u that is available to create and maintain internal order; thus,

$$u = \beta m , \qquad\qquad (XII.1)$$

where $\beta$ is a constant of proportionality. The maximization of the mass of the individual organism is then equivalent to maximizing the amount of energy available for creating and maintaining thermodynamic order. An organism expends energy by converting carbon from an essentially reduced state to $CO_2$. The total rate that energy is expended b, the amount of behavior, equals the rate that energy is available to create and maintain internal order u plus the time rate of energy expenditure necessary to do mechanical work q; or,

$$u = b - q . \qquad (XII.2)$$

This energy expenditure u is to be maximized where u is to be regarded as a dependent function of everything happening to the organism.

It may be noted that the ecological efficiency of the organism, m/b, should not be maximized as a dependent function, since it is, in fact, a ratio of dependent functions and such a ratio should never be maximized or minimized (see an analogous discussion in Chapter II concerning the optimization of the performance of a heat engine and the concept of efficiency). However, once m has been maximized, it may be divided by b, which then expresses the precise meaning of the idea that the ecological efficiency of an organism tends toward a maximum.

### Energy Expended to Do Mechanical Work

If w is the time rate that work is done, then a partial efficiency $\eta$ may be defined by

$$q = w/\eta . \qquad (XII.3)$$

The partial efficiency $\eta$ is the efficiency of the organism in converting chemical energy, beyond the requirements of maintaining internal order, into mechanical work. The energy expenditure to maintain order, Equation XII.2, may then be written

$$u = b - w/\eta . \qquad (XII.4)$$

The partial efficiency is itself a function of the time rate that mechanical work is done. It may be assumed that the partial efficiency will satisfy the following requirements: When the rate of working goes to zero, $w \to 0$, then the partial efficiency becomes a maximum, $\eta \to \eta_0$. When the rate of working approaches an upper limit, $w \to w_m$, then the partial efficiency goes to a minimum, $\eta \to \eta_1$, where $\eta_o > > \eta_1$. For illustrative purposes these requirements may be satisfied by a linear approximation,

$$\eta = \eta_o - (\eta_o - \eta_1) w/w_m . \tag{XII.5}$$

From Equations XII.3 and XII.5 the energy that has to be expended per unit time, in addition to u, to accomplish a rate of doing work w is then given by

$$q = w[\eta_o - (\eta_o - \eta_1) w/w_m]^{-1} . \tag{XII.6}$$

This result is shown in Figure 32.

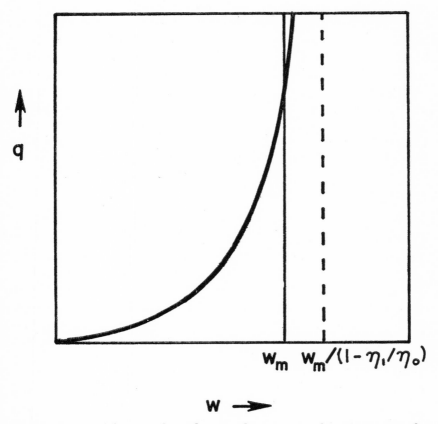

$$w_m \qquad w_m/(1 - \eta_1/\eta_o)$$

$$w \longrightarrow$$

Figure 32. A qualitative plot indicating the time rate that energy must be expended (in addition to the rate that energy must be expended to maintain internal order, u) to do work, q, as a function of the time rate of doing work, w, Equation XII.6.

## External and Internal Work

When no work is done $q = 0$ and Equation XII.2 gives $u = b$. Thus, u represents a basal or resting metabolic rate (the basal metabolic rate referred to here is not the mass specific measure). Since an organism can expend various amounts of energy while involved in different activities that do not involve any work done on the environment, and since over short periods of time u may be taken as a constant, w, as defined by Equations XII.3 and XII.2, must involve internal work as well as external work. For example, a man walking (neglecting frictional losses between his feet and the ground) or standing at attention does no work on the environment; but according to Equations XII.3 and XII.2 he must be doing work, since he is expending energy at a rate greater than the basal rate. To make the situation clearer, the same problem also arises in the case of an automobile moving slowly along a smooth level road. No work is done by the automobile on the environment, neglecting the frictional losses due to air resistance and between the tires and the road. The gasoline engine in the automobile converts chemical energy into mechanical work which is then used internally to overcome internal friction. It, thus, becomes useful to differentiate between internal and external work where

$$w = w_e + w_i , \qquad (XII.7)$$

where $w_e$ is the time rate of doing external mechanical work on the environment and $w_i$ is the time rate of doing work internally.

Internal work in inorganic systems is almost always work expended to overcome internal friction. The internal work performed by carbon-based life, however, not only involves work to overcome internal friction, it also involves work to maintain a particular structural conformation or state. Animals, by changing the rigidity or tetanus in muscles, are capable of changing their structural state and are, thereby, capable of movement and doing external mechanical work. The amount of tetanus in a muscle, as well as the rate of doing mechanical work on the environment and to overcome internal friction, determines the rate that energy must be expended. A single muscle contracting will develop a tension T equal to the force it can exert F. If the single muscle contracts with a velocity v, it will deliver work at the rate

$$w_e = Fv = Tv \ . \qquad\qquad (XII.8)$$

The tension is never quite zero, so that it requires work to extend the muscle once it has contracted. The single muscle must not only expend energy at the rate indicated by Equation XII.8, but it must, in addition, expend energy to overcome internal friction in the muscle itself, $w_f$, and to maintain the tension T, even if no external work were ever done, $w_t$. The rate internal work is done then becomes

$$w_i = w_f + w_t \ . \qquad\qquad (XII.9)$$

To a first approximation the rate of energy expended in internal friction, $w_f$, may be taken as proportional to the tension developed and the rate of doing work,

$$w_f = \gamma \, v^2 T \ , \qquad\qquad (XII.10)$$

where $\gamma$ is a constant of proportionality.

To a first approximation the rate that energy must be expended in a single muscle to maintain the tension T, apart from any external work done, $w_t$ may be taken as proportional to the tension,

$$w_t = v^* T \ , \qquad\qquad (XII.11)$$

where $v^*$ is a positive constant of proportionality that may itself be a function of T. The total rate of work by a single muscle from Equations XII.8 through XII.11 then becomes

$$w = (v^* + v + \gamma \, v^2) T \ . \qquad\qquad (XII.12)$$

If the whole complex of muscles and skeletal structures are considered, the rate that external work is done becomes a complicated function of the geometry and the rate of contraction of each muscle. Nevertheless, for n muscles it is possible to write

$$w_e = \sum_{i=1}^{n} \varepsilon_i v_i T_i \ , \qquad\qquad (XII.13)$$

where $\varepsilon_i < 1$ and the values of $\varepsilon_i$ and $v_i$ depend upon the geometry. The rate that internal work is done, neglecting friction in skeletal joints, becomes simply additive; or

$$w_i = \sum_{i=1}^{n} (\gamma v^2_i + v^*_i) \ T_i \ . \tag{XII.14}$$

Since muscles contract and relax periodically during most activities, the effective values of the work done are time average values obtained by integrating Equations XII.8 through XII.14 over suitable time periods.

### Available Chemical Energy

When an organism ingests food a certain amount of this food is assimilated and is made available to the organism in the form of chemical energy. On the average the amount of energy made available by the ingestion of food must equal the energy expended. This means that the total amount of behavior, i.e. the total energy expended, by the organism must equal the total amount of food energy made available over any long time period.

The relationship between the amount of food ingested and the energy made available is not a linear relationship. The amount of energy made available depends upon the efficiency of assimilation; losses include all of the processes of mastication, digestion and storage. If r is the time rate that energy is made available and f is the time rate that food energy is ingested, then

$$r = \alpha f , \tag{XII.15}$$

where $\alpha$ is the efficiency of assimilation which is itself a function of f. In particular, the following criteria should be approximately satisfied: If $f \to 0$, then $\alpha \to \alpha_m$ where $\alpha_m$ is a maximum. And if $f \to \infty$, then $\alpha \to 0$ and $r \to r_m$ where $r_m$ is an upper limit to the rate that energy can be made available. For illustrative purposes, or to a first approximation, the relationship between the time rate that energy is made available, r, and the time rate that food is ingested, f, may then be taken as

$$r = \alpha_m f / (1 + \alpha_m f / r_m) . \tag{XII.16}$$

This function is shown in Figure 33 where for a suitably long time average with no work done $u = r$.

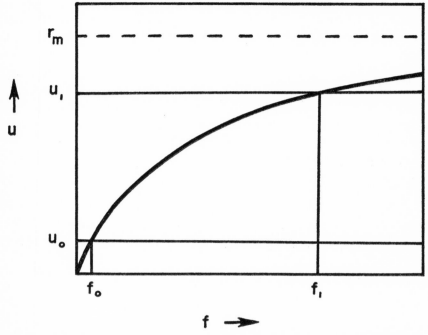

Figure 33. Illustration of the time rate that energy is made available to maintain internal order, u, as a function of the time rate that food energy is ingested, f, Equation XII.19, when no work is done.

The time average amount of behavior $<b>$ must equal r, assuming r to be a steady state average. In particular, if the day is divided up into n different activities with amounts of behavior $b_i$ and each lasting for a period of time $\triangle t_i$, where $\sum_{i=1}^{n} \triangle t_i = t_o$ where $t_o$ is twenty-four hours, then

$$r = <b> = t_o^{-1} \sum_{i=1}^{n} b_i \triangle t_i. \qquad (XII.17)$$

If the amount of behavior is specified as a continuous function of the time then the summation may be replaced by an integral,

$$r = <b> = t_o^{-1} \int_0^{t_o} b \, dt. \qquad (XII.18)$$

## Behavior when Food Energy Is Limited and No Work Is Done

There are an infinite variety of circumstances to which an organism may be required to adjust. Predicting behavior can become as difficult as the complexity of these circumstances. Only the simplest situations will be considered here. The present section treats the situation in which an animal is presented with a limited amount of food each day. Apart from the limited amount of food calories, it is assumed that all compounds essential for life are supplied in excess. It is further assumed that the animal does no work.

Since the animal is engaged in only the one activity of eating (where the work required to eat is included implicitly in the efficiency of assimilation $\alpha$), the time rate that energy is made available r may be set equal to the amount of behavior b. From Equations XII.2 and XII.16, where $q = 0$ since no work is done, the rate that energy is expended to maintain internal order becomes

$$u = \alpha_m f / (1 + \alpha_m f / r_m) . \qquad (\text{XII.19})$$

It is assumed that the rate of expenditure of energy to maintain internal order must be greater than some minimum value $u_o$ to avoid death. For organisms not completely dormant $u_o > 0$. It will also be assumed that u is bounded above by $u_1$, there being a limit to the size of an organism imposed by the size of the skeletal structure and other limitations; thus,

$$u_o \leqq u \leqq u_1 . \qquad (\text{XII.20})$$

To find the rate that an animal will ingest food energy f it is necessary to maximize u, Equation XII.19, noting the additional restrictions Equation XII.20. Malfunctions of perception, learning, etc., are not being considered here, the animal being assumed to be adequately adapted to its niche. The range of choices that is in principle available to the animal is from $f = f_o$, where $u_o = u(f_o)$ is the lower bound on u, to the limited amount of food energy actually provided f', where it is assumed that $f' \leqq f_1$ where $u_1 = u(f_1)$ is the upper bound on u, as shown in

Figure 33. When the amount of food presented to the animal exceeds $f_1$, the amount ingested remains at $f_1$, or less, since under these circumstances $f_1$ cannot be exceeded without $u_1$ being exceeded. Since the function $u(f)$ given by Equation XII.19 has no proper maximum, the maximum value of u is given at the upper bound of the interval of choice that is available. This upper bound occurs when all of the food presented is eaten, i.e. when $f = f'$ where $f' \leqq f_1$.

The situation being considered in the present section involves long time averages and is a steady state situation. The quantity u is slow to change with time and is somewhat resistant to change. It will probably require patience to verify the conclusions of this section in the laboratory, perhaps a different animal being needed for each point on the experimental curve showing $u = u(f)$.

Since the energy available to maintain internal order is proportional to the body mass of the organism, Equation XII.1, the relationship between u and f, Equation XII.19, may be interpreted as a relationship between the body mass m and the rate of ingestion of food f when no work is being done.

## Behavior with Work

From Equations XII.2, XII.6 and XII.16 with $b = r$ for the long time average the rate that internal energy is available to maintain order may be written as

$$u = \frac{\alpha_m f}{1 + \alpha_m f/r_m} - \frac{w}{\eta_o - (\eta_o - \eta_1)w/w_m} , \qquad (XII.21)$$

where $u_o \leqq u \leqq u_1$. Maximizing Equation XII.21, it is apparent that animals will behave to maximize the rate at which food energy is ingested and minimize the rate at which they do work. If an infinite amount of food is always available and no work is required, an animal will do no work and will tend to eat the maximum amount of food $f_1$ corresponding to $u_1$ the upper bound on the rate that internal energy can be made available for maintaining internal order. These results appear to be intuitively correct.

## *Case when Food Energy Is Provided as a Function of Work*

Circumstances usually require work to be done in order for food to become available. For example, a cow must not only do work to chew the grass it eats, it must also do work in order to stand and to walk from place to place as it grazes. The time rate that food energy is made available f then becomes a function of the time rate that work is done. The case for which the rate at which food is supplied is fixed independent of the rate of doing work has already been treated in the previous section. The next simplest case occurs when the rate that food energy is made available is directly proportional to the rate that work is done,

$$f = kw \,, \qquad (XII.22)$$

where k is a constant of proportionality.

The quantity k measures the efficiency of gathering food energy. It measures the amount of food energy obtained over the amount of work that has to be done to obtain the food energy. The quantity k must always be a large quantity greater than one, since an animal has the ability of converting only a fraction of the food energy gathered into useful work, some of the work being in turn required to gather food. In the next section it will be shown that $k > (\alpha_m \eta_o)^{-1}$ where $\alpha_m$ and $\eta_o$, defined by Equations XII.16 and XII.6, respectively, are both less than unity. A cow kept in a barn and supplied with hay will be operating under conditions for which k may be extremely large. On the other hand, a cow grazing on very poor land will be faced with a situation in which k becomes very small (perhaps as low as 5).

Substituting Equation XII.22 into XII.21, an expression for the average time rate that internal energy can be expended to maintain internal order u is obtained as a function of the rate that food energy is ingested f; thus,

$$u = \frac{\alpha_m f}{1 + \alpha_m f / r_m} - \frac{f/k}{\eta_o - (\eta_o - \eta_1) f / k w_m} \,, \qquad (XII.23)$$

where $u_o \leqq u \leqq u_1$ . This function is shown in Figure 34 for three choices of k.

Maximizing u by differentiating Equation XII.23 with respect to f and setting the result equal to zero yields the time

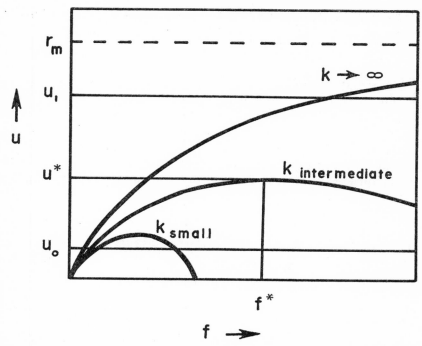

Figure 34. A sketch indicating the time rate that energy is expended to maintain internal order, u, as a function of the time rate food energy is ingested, f, for three values of the efficiency of food gathering, k, Equations XII.23 and XII.24.

average choice for the rate that food energy will be consumed; or

$$f^* = a(a\text{-}1)/\alpha_m [(\eta_o - \eta_1)/w_m + a/r_m] , \quad \text{(XII.24)}$$

where

$$a = (\eta_o \alpha_m k)^{\frac{1}{2}} . \quad \text{(XII.25)}$$

It may be noted that since f must be greater than zero for survival, then $a > 1$, which from Equation XII.25 means that

$$k > (\alpha_m \eta_o)^{-1} , \quad \text{(XII.26)}$$

and $k > 1$, since both $\alpha_m$ and $\eta_o$ are less then unity. If it is assumed that k is very large so that $a \gg 1$, Equation XII.24 yields the approximation

$$f^* \approx r_m (\eta_o k/\alpha_m)^{\frac{1}{2}} . \quad \text{(XII.27)}$$

Assuming $\eta_o$ and $\alpha_m$ are of the same order of magnitude, the order of magnitude of f* becomes, from Equation XII.27,

$$f^* \sim r_m k^{\frac{1}{2}} . \qquad (\text{XII.28})$$

Since the process of food gathering always requires some work, k can never become infinite; consequently, it may be seen from Equation XII.28 that f* will always be limited, and u* will always be limited to values below $r_m$, according to Equation XII.23, even if the upper bound at $u = u_1$ were not assumed.

Substituting Equation XII.24 into Equation XII.23, the maximum time average rate that energy can be expended to maintain internal order is given by

$$u^* = (a - 1)^2 / [(\eta_o - \eta_1)/w_m + a^2/r_m] . \qquad (\text{XII.29})$$

This maximum value of u is shown in Figure 35 as a function of k. Since life cannot be sustained for u less than $u_o$, the value of k must always be larger than $k_o$, where $u^*(k_o) = u_o$.

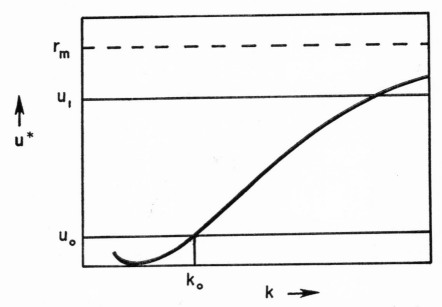

Figure 35. A qualitative plot of the maximum average time rate that energy can be expended to maintain internal order, u*, as a function of the efficiency of food gathering, k, Equations XII.29 and XII.25.

The time average rate that an animal will do work under the present circumstances may be obtained by substituting Equation XII.24 into XII.22; thus,

$$w^* = \eta_0(1 - a^{-1})/[(\eta_0 - \eta_1)/w_m + a/r_m], \quad (XII.30)$$

where a is defined by Equation XII.25. The work done $w^*$, Equation XII.30, is shown in Figure 36 as a function of k. As k

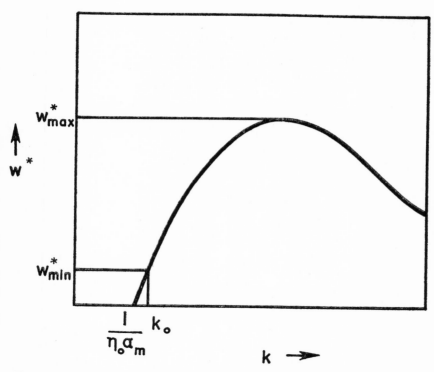

Figure 36. A sketch showing how the average time rate that work is done, $w^*$, varies as a function of the efficiency of food gathering, k, Equations XII.30 and XII.25.

becomes small and the work done becomes less productive in producing food, the time average rate of doing work decreases. According to Equation XII.30 (neglecting for the moment the fact that $u \geqq u_0$), when k is made small enough a point is reached where no work is done, $w^* = 0$. This occurs when $a = 1$, or from Equation XII.25, when $k = (\alpha_m \eta_0)^{-1}$. In this case the food eaten would not be sufficient to do the work when the efficiency of assimilation is at best $\alpha_m$ and the efficiency of working is at best $\eta_0$, since it may be seen from Equation XII.29 that $u = 0$ and no energy would be left at all to maintain internal order.

Since u can never be less than the minimum value $u_0$, the requirement specified by Equation XII.22 indicates that a minimum amount of work will always have to be done to obtain the minimum amount of food to sustain life. This minimum, $w^*_{min}$, may be obtained by finding $k_0$ from Equation XII.29 where $u_0 = u^*(k_0)$ and substituting into Equation XII.30. It is apparent that an animal will perish while still working and striving to gain sufficient food when k is reduced to just below the minimum value for survival.

As k is increased and the work becomes more productive in producing food, the time average rate of doing work will increase to a maximum, $w^*_{max}$. Differentiating Equation XII.30 with respect to a (since, according to Equation XII.25, $k = a^2 / \alpha_m \eta_0$) and setting the result equal to zero, the maximum rate of working will be obtained when

$$a = 1 + [1 + (\eta_0 - \eta_1) r_m / w_m]^{\frac{1}{2}} . \qquad (XII.31)$$

For $r_m >> w_m$ and $\eta_1 << \eta_0$, this value of a is such that $a \gtrsim (\eta_0 r_m / w_m)^{\frac{1}{2}}$. From Equations XII.31 and XII.25 the value of k that will produce the maximum rate of doing work, $w^*_{max}$, is then

$$k = (\alpha_m \eta_0)^{-1} \{ 1 + [1 + (\eta_0 - \eta_1) r_m / w_m]^{\frac{1}{2}} \}^2 , \qquad (XII.32)$$

which for $r_m >> w_m$ is such that $k \gtrsim r_m / w_m \alpha_m$. The maximum time rate of working may then be obtained by substituting Equation XII.31 into Equation XII.30,

$$w^*_{max} = \eta_0 r_m \{ 1 + [1 + (\eta_0 - \eta_1) r_m / w_m]^{\frac{1}{2}} \}^{-2} , \qquad (XII.33)$$

which for $r_m >> w_m$ is such that $w^*_{max} \lesssim w_m$ where $w_m$ is a constant that serves as an upper bound that may never be attainable in actuality.

As k is increased further and less work is needed to supply the food energy requirements of the animal, the time average rate of working decreases asymptotically to zero, as shown in Figure 36. Since it is permissible to allow k to become very large, the upper bound $u \leqq u_1$ does not keep the rate of doing work from going to zero.

Since the total work done by an animal may be proportional to the useful work done by the animal, the efficiency defined by the ratio of total work done to the total consumption of food energy may be of interest. From Equation XII.30 this efficiency is given by $w^*/f^* = 1/k$. This efficiency is a maximum when k is a minimum, or when $k = k_o$ where $u^*(k_o) = u_o$. Thus, according to the present simple theory, the greatest amount of work that can be done for a given amount of food is done when an animal ingests food at the minimum rate for survival. This result may approximate the true situation where a maximum efficiency of work to food may occur for a rate of food consumption slightly above the minimum survival rate.

### Time Dependent Behavior under Deprivation with no Work Done

In the previous section it was found that the time average or steady state behavior of an organism with respect to the ingestion of food energy and the performance of work could be predicted, in principle, with little difficulty. Problems involving the time explicitly are much more difficult.

The rate that energy is available to maintain internal order is dependent upon the ingestion of vital compounds in addition to energy yielding compounds. Such compounds as oxygen, water and various salts are essential for an animal to carry on metabolism. The rate that energy can be expended, the amount of behavior b, depends upon the rate that these compounds are used in the body. If the time rate that the $i$th compound is used up in the body is $c_i$ (perhaps, in units of grams per day), then the amount of behavior b becomes

$$b = b(c_1, c_2, \ldots, c_n) , \qquad (XII.34)$$

where there are n vital compounds. Since life ceases whenever any one of the vital compounds becomes exhausted, the condition $b \to 0$ as $c_i \to 0$ must be valid for each value of i. In general it may be assumed that there will be some value of $c_i$ for which b can be a maximum. This means that b cannot attain its full value if $c_i$ is too small and that beyond a certain point if $c_i$ becomes too large to interfere with the metabolic processes and will again cause b to decrease. Thus, for illustrative purposes or to a crude first approximation it may be assumed that the effect of each of the vital compounds is independent (which cannot be strictly true), and b may be taken as

$$b = b_m \prod_{i=1}^{n} g_i(c_i) \, , \qquad \qquad (XII.35)$$

where $b_m$ is an upper limit for b and each of the g's satisfies the following conditions:

$$
\begin{array}{lll}
g(c) \to 0 & \text{for} & c \to 0 \, , \\
g(c) \leq 1 & \text{for} & \text{all values of c} \\
g(c_m) = 1 & \text{for} & g(c_m) \text{ a maximum .}
\end{array}
$$
$$ (XII.36) $$

Again for illustrative purposes, or to a crude first approximation, the following simple choice for g(c) will be considered:

$$g(c) = c(2c_m - c)/c_m^2 \, . \qquad \qquad (XII.37)$$

This function may be compared with the choice of function for the case of food energy, Equation XII.16.

An animal has the capacity to store vital compounds in its body. As the compounds are used up the amount stored decreases with time. For illustrative purposes, it will be assumed that a vital compound is used up at a constant steady rate, which depends upon the rate the animal expends energy, until the compound is almost entirely gone. From that time on it will be assumed that the stored vital compound decreases exponentially with time. In particular if K is the amount of vital compound then

$$K = (c_0/\kappa) \begin{cases} 1 - \kappa(t - t^*) & \text{for} \quad 0 \leq t \leq t^* \, , \\ e^{-\kappa(t - t^*)} & \text{for} \quad t^* \leq t \, , \end{cases} \qquad (XII.38)$$

where K, as well as $c = -\,dk/dt$, are assumed to be continuous at $t = t^*$, $c_o$ is the constant rate that the compound is used up, $t^*$ is the time when the amount stored is sufficiently small that an exponential decrease commences, and $\kappa$ is the reciprocal time constant which measures the time it takes the compound to decrease to $1/e$ its value at $t = t^*$. The amount stored at the time $t = 0$ is $K_m = (c_o/\kappa)(1 + \kappa\,t^*)$. It will be assumed that $t = 0$ refers to a time when ingestion of the vital compound occurs. It will be assumed that a maximum amount of the compound is always ingested at this time $t = 0$, so that $K_m$ represents the maximum amount of the vital compound that can be stored. The time rate that the vital compound is used in the body is obtained by differentiating Equation XII.38, or

$$c = c_o \begin{cases} 1 & \text{for} \quad 0 \leqq t \leqq t^*\,, \\ e^{-\kappa(t-t^*)} & \text{for} \quad t^* \leqq t\,. \end{cases} \qquad \text{(XII.39)}$$

These results, Equations XII.38 and XII.39 are shown in Figure 37 for two possible values of $c_o$.

The maximum amount of vital compound that can be stored, $K_m$, will, quite naturally, be a function of the mass of the animal or of the time rate that energy can be expended to maintain internal order, u. The present section is concerned with short time phenomena, so that the body mass is no longer a variable, and $K_m$ may be regarded as a constant. It is also possible to consider situations in which an animal does not replenish its supply of a vital compound to the maximum amount; but such situations are beyond the present simple considerations. Similarly, it might be possible to consider excess amounts of a vital compound that exceed $K_m$, such as occurs in hyperoxygenation produced by breathing heavily; but again such phenomena are beyond the scope of the present discussion.

### Expectation Energy and the Need State

What an animal does at any particular time t can only affect the future. Since an organism behaves such as to maximize the energy available to create and maintain internal order, it must behave so as to maximize the future energy available. In particular, $V(t)$, the expectation energy at time t is the total energy

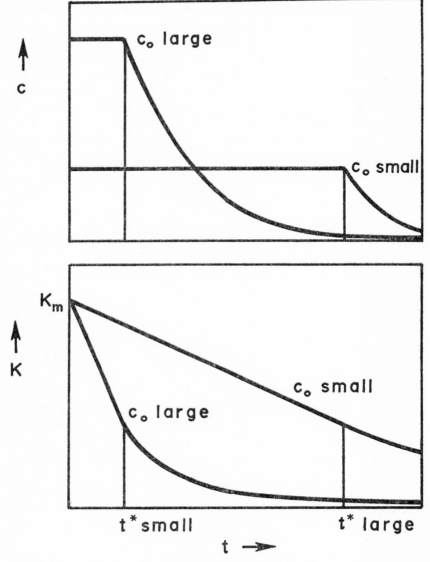

Figure 37. Sketch of the assumed decrease in a vital compound stored in the body, K, and the rate of use of the compound, c, with time of deprivation, t, for two values of $c_o$, Equations XII.38 and XII.39.

expected at the time t to be available to create and maintain internal order in the future; thus,

$$V(t) = \int_t^{\infty} e^{-(t'-t)/\tau} u(t') \, dt', \qquad (XII.40)$$

where $u(t')$ is the rate that energy is expected to be available in the future at time $t'$, the exponential factor, which reduces the integrand to $1/e$ its value when $t' - t = \tau$, is a measure of the expectation of catastrophe such as drought, earthquake, death by predation or disease, etc. (*cf.* Equation V.4). The value of $\tau$ should be comparable to the life span of the individual organism. For the usual situation $u(t')$ will be a rapidly decreasing function (for example, after being deprived of oxygen), so that the exponential factor may be set equal to unity, usually.

While it is possible to predict time dependent behavior by maximizing the expectation energy defined by Equation XII.40, it is intuitively more pleasing to define a need function as

$$N(t) \equiv V_m - V(t), \qquad (XII.41)$$

where $V_m$ is a constant independent of time which measures an ideal maximum expectation energy under the conditions of no deprivations. In particular, if $u = u_1$ the maximum of u, Equation XII.40 yields upon integration

$$V_m = \tau u_1. \qquad (XII.42)$$

The need function measures the failure of the expectation energy to reach the ideal limit. Predicting behavior by maximizing the expectation energy is now seen to be equivalent to minimizing the need function.

Since $u(t')$ can represent the energy available to maintain order under the simultaneous deprivation of different vital compounds ingested at different times in the past, the expectation energy $V(t)$ becomes a function of the various deprivation times. The need function may then be written

$$N = N(t; t_1, t_2, \ldots, t_n), \qquad (XII.43)$$

where t is the time and the times of deprivation of the n vital compounds are $t - t_1$, $t - t_2$, ..., $t - t_n$. It might seem strange that there is only one need function to represent all needs, but there is basically only one fundamental need, the need to create and maintain internal order. Individual needs arise only insofar as they affect the fundamental need.

It is possible to define individual needs, but such individual needs will not be strictly independent. For example, the deprivation of water will affect the need for food and salt. The need for a particular vital compound when an animal is in a particular state may be defined as the difference between the need state if the vital compound is not ingested and the need state if the vital compound were to be ingested. The need for the $j$th vital compound is defined by

$$N_j = N(t; t_1, t_2, \ldots, t_{j-1}, t_j, t_{j+1}, \ldots, t_n) \\ - N(t; t_1, t_2, \ldots, t_{j-1}, t, t_{j+1}, \ldots, t_n), \quad (XII.44)$$

where the time of deprivation of the $j$th compound becomes zero for $t_j = t$. An animal will tend to behave such as to satisfy the greatest of its individual needs, since satisfying the greatest individual need will reduce the need state the most.

Since an animal can generally behave such as to satisfy only one need at a time, the single function, Equation XII.43, appears to be sufficient for predicting.

### Motivation and Need

Before an animal will behave so as to replenish the loss of a vital compound it must have internal sensory mechanisms that make the animal aware of the deficiency in the vital compound. The image or perception of the internal state implied by the sensations reported by such sensory mechanisms then establishes a motivational state. The motivational state may then be regarded as an internal state that is one of the immediate causes of directed behavior. Since an organism can only behave in one direction at a time, in general, there is need to define only one function or state as the motivational state. Individual motivations, the desire to ingest a particular vital compound, for example, may be defined as the difference in the motivational state just before and just after having satisfied the need, such as ingestion of a particular vital compound.

Since an organism tends to behave such as to maximize its internal order, there must, in general, be an excellent correspondence between the need state and the motivational state. Since an organism is equipped to survive adequately only in the habitat in which it developed evolutionarily, it is only within the context of such a natural habitat that the close correspondence between the need state and the motivational state can be expected. Laboratory situations in which an animal will not behave such as to minimize its actual need may be easily devised. For example, most animals, being unaware of the vital role of particular vitamins will perish if the vitamins are included in food that the animal rejects because of a slightly strange or unnatural flavor. There is no motivation for ingesting particular vitamins. Similarly, a stimulation of the sensation of dryness at the back of the throat can lead an animal to drink enough water to drown internally. A high motivation to drink water does not reflect the true need in this case. Although such laboratory experiments may be of interest when investigating the range and accuracy of internal perceptions, they are of little interest when considering the behavior of animals in their natural habitats playing their natural roles.

Under ordinary natural circumstances the motivational state may, thus, be equated to the need state,

$$\Phi = N = V_m - V. \qquad (XII.45)$$

This equality may also be interpreted as the condition for an ideal animal that is ideally equipped to recognize all of its needs. Or it might be equally interpreted as a condition that is satisfied for only a limited variety of circumstances that might be regarded as ideal circumstances. The ideal circumstances must closely approximate the actual circumstances that occur in the animal's natural environment.

### Need State under Deprivation of a Single Compound with No Work Done

When no work is done the need state, as defined by Equations XII.41 and XII.40, is entirely specified when the stored amounts of each of the vital compounds are known. For deprivation of a single compound the amount of behavior, b, according to the approximations, Equations XII.35, XII.37 and XII.39, becomes

$$b = b_m \tau \begin{cases} 2 - \tau & \text{for} \quad t' \leqq t^* , \\ 2e^{-\kappa(t - t^*)} - e^{-2\kappa(t - t^*)} & \text{for} \quad t^* \geqq t' , \end{cases} \tag{XII.46}$$

where t is the time of deprivation, $\tau$ is a constant,

$$\tau = c_o / c_m , \tag{XII.47}$$

and $t^*$ is defined by Equation XII.39. Since no work is being done, $c_o$ represents the depletion rate corresponding to the basal metabolic rate. The constant $c_m$ represents the maximum possible depletion rate.

The expectation energy, Equation XII.40, for no work done, or $u = b$, Equation XII.2, ignoring the slowly varying exponential factor, becomes upon integration

$$V(t) = \frac{b_m \tau}{2\kappa} \begin{cases} 4 - \tau + 2\kappa(2 - \tau)(t^* - t) & \text{for} \quad t \leqq t^* , \\ 4e^{-\kappa(t - t^*)} - e^{-2\kappa(t - t^*)} & \text{for} \quad t^* \leqq t . \end{cases} \tag{XII.48}$$

The individual need for the compound from Equation XII.44 and Equation XII.41 becomes

$$N_1(t) = N(t;0) - N(t;t) = -V(t) + V(0) ; \tag{XII.49}$$

or

$$N_1(t) = \frac{b_m \tau}{2\kappa} \begin{cases} 2\kappa(2 - \tau)t & \text{for} \quad t \leqq t^* , \\ 4 - \tau + 2\kappa(2 - \tau)t^* - 4e^{-\kappa(t - t^*)} + e^{-2\kappa(t - t^*)} & \\ & \text{for} \quad t^* \leqq t . \end{cases} \tag{XII.50}$$

This individual need function is sketched in Figure 38 as a function of the time deprivation t for two values of $b_m$.

The maximum possible need occurs when the minimum metabolic rate is reached just prior to death. From Equation XII.35 and XII.37 death will occur when $t = t_o$ where $b = u_o = b_m c(t_o)$ $[2c_m - c(t_o)]/c_m^2$. To a first approximation, however, $t_o \to \infty$, so that the maximum need from Equation XII.50 becomes

$$N_1(\infty) = b_m \tau [(4 - \tau)/2\kappa + (2 - \tau)t^*] . \tag{XII.51}$$

Since $b_m$ varies directly as the mass of the animal, the need also varies directly as the mass. It may also be noted from Equa-

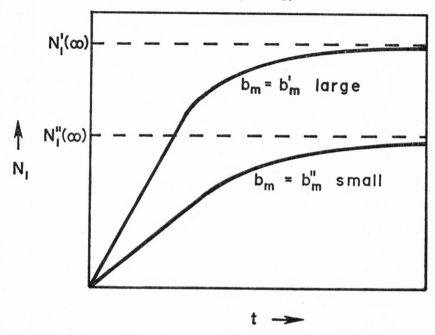

Figure 38. Sketch of the need, $N_1$, for a single compound, Equation XII.49, as a function of the time of deprivation, t. The upper curve is for a more massive animal.

tion XII.50 that the time rate of increase of the need also increases with the mass of the animal. The need of an elephant as compared with the need of a mouse will be, thus, approximately equal to the ratio of their weights. A mass specific need could be defined, but the role of mass is too important to hide in such a manner. An animal that increases its mass and the rate at which energy is expended to maintain internal order, will also increase its need state, according to Equation XII.50, when deprived. It might, thus, seem contradictory to maximize u and at the same time minimize $N(t)$. This, however, is not the case; the fundamental idea was to maximize u. In fact, the minimization of $N(t)$ was obtained by maximizing u. The apparent difficulty here arises from the fact that the integration of Equation XII.41 to obtain the result Equation XII.50 was carried out assuming that $b_m$ (and, thus, the steady state mass of the animal) did not change with time. Consequently, $N(t)$, as given by Equation XII.50 (but not necessarily as given by the general definition, Equation XII.41), applies only to short time periods during which $b_m$ does not change appreciably with time.

An animal to maintain its need at a minimum if it has need for only one vital compound, as considered in the present section, will ingest the compound continuously. In this fashion it may be seen, by letting $t = 0$ in Equation XII.50, that the animal will be able to maintain its individual need for this compound always at zero. Most organisms have need of many different compounds, so that their actual behavior does not degenerate to such a simple pattern.

### Need for Experiments

The present theory indicates that there is a very great need for experiments that measure the energies involved in behavior. Essentially nothing has been done in this important area of experimental research as yet. No experimenter has as yet measured motivational strengths in energy units as suggested by the present theory. The rate that a rat presses a lever in a Skinner box (Heron and Skinner, 1967) may be proportional to the rate of energy expenditure, but no one has as yet determined the proportionality constant. Activity measurements using an activity cage are similarly never converted to fundamental units of the rate of energy expenditure. Because of the fundamental role of energy in science, it may be safely stated that the psychology of motivation will develop into a true quantitative science only after the appropriate experiments (such as indicated in the present Chapter) involving energy have been performed. In the literature there are some real attempts to put motivation on a scientific quantitative scale (e.g. Birch and Veroff, 1966), but unfortunately the greatest volume of literature indicates little recognition of the real problems and little desire to resolve the problems by experiment (e.g. Weiner, 1972).

# REFERENCES

Abeloos, M.: Recherches expérimentales sur la croissance et la régénération chez les planaires. *Bull Biol Fr Belg, 64:*1, 1930.

Abramowitz, M., and Stegun, I. A. (Eds.): *Handbook of Mathematical Functions.* New York, Dover, 1965, p. 297.

Allee, W. C.: *The Social Life of Animals,* revised ed. Boston, Beacon, 1938 (reprinted in 1958).

Alman, P. L., and Dittmer, D. S. (Eds.): *Biology Data Book.* Federation of American Societies for Experimental Biology, Washington, D.C., 1964.

Alpatov, W. W., and Pearl, R.: Influence of temperature during the larval period and adult life on the duration of the life of the imago of *Drosophila melanogaster. Am Nat, 63:*37, 1929.

Andrew, W.: *The Anatomy of Aging in Man and Animals.* New York, Grune & Stratton, 1971.

Andrewartha, H. G.: *Introduction to the Study of Animal Populations.* Chicago, University of Chicago Press, 1961a.

———: *Introduction to the Study of Animal Populations.* Chicago, University of Chicago Press, 1961b, Ch. IX.

Ardrey, R.: *The Territorial Imperative.* New York, Dell, 1966.

Asimov, I.: *Life and Energy.* New York, Doubleday, 1962, p. 302.

Ayala, F. J.: Competition between species: frequency dependence. *Science, 171:*820, 1971.

Ayres, E., and Scarlott, A.: *Energy Sources—The Wealth of the World.* New York, McGraw-Hill, 1952, p. 244.

Barghoorn, E. S., and Tyler, S. A.: Microorganisms of middle precambrian age from the Animikie series, Ontario, Canada. In Mamikunian, G., and Briggs, M. H. (Eds.): *Current Aspects of Exobiology.* Oxford, Pergamon, 1965, p. 93.

Bartholomew, G. A., Howell, T. R., and Cade, T. J.: Torpidity in the white-throated swift, anna hummingbird, and poor-will. *Condor, 59:*145, 1957.

Bender, M. L.: Carbon isotope fractionation. In Fairbridge, R. W. (Ed.): *The Encyclopedia of Geochemistry and Environmental Sciences.* New York, Van Nostrand Reinhold, 1972, p. 133.

Bendixson, I.: Sur les courbes définies par des équations différentielles. *Acta Math, 24:*1, 1901.

Berg, B. N., and Harmison, C. R.: Growth, disease, and aging in the rat. *J Gerontol, 12*:370, 1957.

Birch, D., and Veroff, J.: *Motivation: A Study of Action.* Belmont, California, Brooks/Cole, 1966.

Bourlière, F.: The comparative biology of aging. *J Gerontol, 13*(1):16, 1958.

Broms, A.: *Thus Life Began.* Garden City, New Jersey, Doubleday, 1968.

Burns, R. O.: Genetic variation and gene transfer. In Joklik, W. K., and Smith, D. T. (Eds.): *Zinsser Microbiology,* 15th ed. New York, Meredith, 1972, Ch. 8, p. 129.

Cairns-Smith, A. G.: *The Life Puzzle.* Toronto, University of Toronto Press, 1971.

Callan, H.: *Ethology and Society.* Oxford, Clarendon, 1970, p. 72.

Campbell, A.: Episomes in evolution. In Smith, H. R. (Ed.): *Evolution of Genetic Systems,* a symposium at Brookhaven National Laboratory, New York, Gordon and Breach, 1972, p. 534.

Carson, R.: *Silent Spring.* Boston, Houghton Mifflin, 1967.

Chalmers, J. A.: *Atmospheric Electricity.* London, Pergamon, 1957, p. 246.

Cellarius, R. A., and Platt, J.: Councils of urgent studies; coordinating councils could focus and legitimize research on solutions of our major crises. *Science, 177*:670, 1972.

Chase, S.: *Men and Machines.* New York, Macmillan, 1929.

Commoner, B.: *The Closing Circle, Nature, Man, and Technology.* New York, Knopf, 1971.

Cousteau, J.-Y., and Diolé, P.: *The Whale, Mighty Monarch of the Sea,* trans. by J. B. Bernard. Garden City, New Jersey, Doubleday, 1972.

Curtis, H. J.: *Biological Mechanisms of Aging.* Springfield, Illinois, Thomas, 1966.

Curtis, R.: Bacterial conjugation. *Annu Rev Microbiol, 23*:69, 1969.

Daniels, F.: *Direct Use of the Sun's Energy.* New Haven, Yale University Press, 1964.

Darwin, C.: *The Origin of Species.* New York, New American Library of World Literature, 1859 (reprinted in 1958).

Davis, H. T.: *Introduction to Nonlinear Differential and Integral Equations.* New York, Dover, 1962.

Dawson, P. S., and King, C. E. (Eds.): *Readings in Population Biology.* Englewood Cliffs, New Jersey, Prentice-Hall, 1971.

de Groot, S. R.: *Thermodynamics of Irreversible Processes.* Amsterdam, North-Holland, 1952.

Dobzhansky, T.: *Genetics and the Origin of Species,* 3rd ed. New York, Columbia University Press, 1951. It contains good bibliography up to 1950.

———: Darwinian evolution and the problem of extraterrestrial life. *Perspect Biol Med, 15*:157, 1972.

————, Hecht, M. K., and Steere, W. C. (Eds.): *Evolutionary Biology.* New York, Meredith, vol. 1, 1967; vol. 2, 1968; vol. 3, 1969.

Dodd, A. P.: *Bull Entomol Res, 27*:503, 1936.

Donnelly, R. J., Herman, R., and Prigogine, I. (Eds.): *Non-equilibrium Thermodynamics, Variational Techniques and Stability,* Chicago, University of Chicago Press, 1966.

Drake, E. T. (Ed.): *Evolution and Environment,* A symposium at Yale University. New Haven, Yale University Press, 1968.

Drake, J. W.: *The Molecular Basis of Mutation.* San Francisco, Holden-Day, 1970.

Ehrensvärd, G.: *Life: Origin and Development.* Chicago, University of Chicago Press, 1962.

Ehrlich, P. R., and Ehrlich, A. H.: *Population, Resources, Environment; Issues in Human Ecology,* 2nd ed. San Francisco, W. H. Freeman, 1972.

Eimbinder, J. (Ed.): *Semiconductor Memories.* New York, Wiley-Interscience, 1971.

Elandt-Johnson, R. C.: *Probability Models and Statistical Methods in Genetics.* New York, John Wiley, 1971.

Ellul, J.: *Technological Society,* trans. by J. Wilkinson. New York, Vintage Books, 1964.

Elton, C.: *Animal Ecology.* London, Sidgwick and Jackson, 1927.

*Eng News Rec,* French Explore sea depths for power and fresh water. *150:* No. 11, 38, 1953.

Epstein, P. S.: *Textbook of Thermodynamics.* New York, Wiley and Sons, 1937.

Feigenbaum, E. A., and Feldman, J. (Eds.): *Computers and Thought.* New York, McGraw-Hill, 1963.

Fink, D. G.: *Computers and the Human Mind.* Garden City, New Jersey, Doubleday, 1966.

Fischman, D. A., and Weinbaum, G.: Hexagonal pattern in cell walls of *Escherichia coli B. Science, 155*:472, 1967.

Fountaine, M. E.: Notes and observations: rapid development of a tropical butterfly. *Entomologist, 71*:90, 1938.

Fox, S. W.: Self-assembly of the protocell from a self-ordered polymer. In Kimball, A. P., and Oro, J. (Eds.): *Prebiotic and Biochemical Evolution.* London, North-Holland, 1971, pp. 8-30.

Gates, D. M.: *Energy Exchange in the Biosphere.* New York, Harper and Row, 1962.

Gause, G. F.: *The Struggle for Existence.* Baltimore, Williams and Williams, 1934.

————, and Witt, A. A.: Behavior of mixed populations and the problem of natural selection. *Am Nat, 69*:596, 1935.

Gelfant, S., and Smith, J. G.: Aging: noncycling cells an explanation. *Science, 178*:357, 1972.

Gilpin, M. E.: Enriched predator-prey systems: theoretical stability. *Science, 177,* 902, 1971.

Gilvarry, J. J.: The possibility of primordial lunar life. In Mamikunian, G., and Briggs, M. H. (Eds): *Current Aspects of Exobiology.* New York, Pergamon, 1965, p. 179.

Glansdorff, P., and Prigogine, I.: *Thermodynamics of Structure, Stability, and Fluctuations.* New York, Wiley-Interscience, 1971.

Goel, N. S., Maitra, S. C., and Montroll, E. W.: On the Volterra and other nonlinear models of interacting populations. *Rev Mod Phys, 43*:231, 1971.

Golde, R. H.: Frequency of occurrence of lightning flashes to the earth. *Q J R Meterol Soc, 71*:89, 1945.

Goldsmith, E., Allen, R., Allaby, M., Davoll, J., and Lawrence, S. (Eds.): *Blueprint for Survival.* Boston, Houghton Mifflin, 1972.

Gompertz, B.: On the nature of the function expressive of the law of human mortality, and on a new mode of determining the value of life contingencies. *Phil Trans, 115*:513, 1825.

Grasselli, A. (Ed.): *Automatic Interpretation and Classification of Images,* a NATO Advanced Study Institute in Pisa in 1968. New York, Academic, 1969.

Gray, T. J., and Gashus, O. K. (Eds.): *Tidal Power,* proceedings of a conference at Halifax, Nova Scotia. New York, Plenum, 1972.

Gutenberg, B.: Seismological and related data. In *American Institute of Physics Handbook,* 2nd ed. New York, McGraw-Hill, 1957, sect. 2j, p. 2-120.

Hammond, A. L.: Geothermal energy: an emerging major resource. *Science, 177*:978, 1972.

Hatsopoulos, G. N., and Keenan, J. H.: *Principles of General Thermodynamics.* New York, John Wiley & Sons, 1965.

Hawkes, J., and Woolley, L. (Eds.): *History of Mankind, Prehistory and the Beginnings of Civilization.* New York, Harper & Row, 1963.

Heinrich, B.: The energetics of the bumblebee. *Sci Am, 228*:96, 1973.

Herbicide Assessment Commission of the AAAS: *War Related Civilian Problems in Indochina,* part 1, Vietnam. Senate Judiciary Committee, Washington, D.C., 1970.

Heron, W. T., and Skinner, B. F.: Changes in hunger during starvation. In Cicala, G. A. (Ed.): *Animal Drives.* New York, Van Nostrand Reinhold, 1967, p. 136.

Herskowitz, G. J. (Ed.): *Computer-Aided Integrated Circuit Design.* New York, McGraw-Hill, 1968.

Higgins, J.: The theory of oscillating reactions. *Ind Eng Chem, 59*:No. 5, 19, 1967.

Hirschfelder, J. O., Curtiss, C. F., and Bird, R. B.: *Molecular Theory of Gases and Liquids.* New York, John Wiley and Sons, 1954, p. 631.

Hodgman, C. D. (Ed.): *Handbook of Chemistry and Physics.* Cleveland, The Chemical Rubber Publishing Co., 1961, p. 1935.

Holling, C. S.: The components of predation as revealed by a study of small-mammal predation of the European pine sawfly. *Can Entomol, 91*: 293, 1959.

———: Some characteristics of simple types of predation and parasitism. *Can Entomol, 91*:385, 1959.

———: The functional response of predators to prey density and its role in mimicry and population regulation. *Mem Entomol Soc Can, 45*:1, 1965.

Ingersoll, L. R., Zobel, O. J., and Ingersoll, A. C.: *Heat Conduction with Engineering, Geological, and Other Applications.* New York, McGraw-Hill, 1948, p. 13.

Jackson, F., and Moore, P.: Possibilities of life on mars. In Mamikunian, G., and Briggs, M. H. (Eds.): *Current Aspects of Exobiology.* New York, Pergamon Press, 1965, p. 243.

Jerome, H.: *Mechanization in Industry.* National Bureau of Economic Research, New York, 1934.

Johnson, H. D., Kintner, L. D., and Kibler, H. H.: Effects of 48 F. (8.9C.) and 83 F. (28.4C.) on longevity and pathology of male rats. *J Gerontol, 18*:29, 1963.

Kaula, W. M.: *An Introduction to Planetary Physics, the Terrestrial Planets.* New York, John Wiley and Sons, 1968a, pp. 253-299.

———: *An Introduction to Planetary Physics, the Terrestrial Planets.* New York, John Wiley and Sons, 1968b, pp. 1 and 110.

Kerner, E. H.: *Gibbs Ensemble; Biological Ensemble.* New York, Gordon & Breach, 1971.

———: A statistical mechanics of interacting biological species. *Bull Math Biophys, 19*:121, 1957.

———: Further considerations on the statistical mechanics of biological associations. *Bull Math Biophys, 21*:217, 1959.

———: On the Volterra-Lotka principle. *Bull Math Biophys, 23*:141, 1961.

Kibler, H. H., and Johnson, H. D.: Metabolic rate and ageing in rats during exposure to cold. *J Gerontol, 16*:13, 1961.

Kinsman, B.: *Wind Waves, Their Generation and Propagation on the Ocean Surface.* Englewood Cliffs, New Jersey, Prentice-Hall, 1965.

Kirkley, D. W.: Determination of the optimum configuration for a Stirling engine. *J Mech Eng Sci, 4*:204, 1962.

Kneese, A. V. and Bower, B. T. (Eds.): *Environmental Quality Analysis,* papers from a Resources for the Future Conference. Baltimore, Johns Hopkins Press, 1972.

Lack, D.: *Darwin's Finches.* New York, Harper & Brother, 1947 (reprinted in 1961).

———: *Ecological Adaptations of Breeding in Birds.* Oxford, Oxford University Press, 1969.

Lave, L. B.: *Technological Change: Its Conception and Measurement.* Englewood Cliffs, New Jersey, Prentice-Hall, 1966.

Leakey, L. S. B.: *Adam's Ancestors.* New York, Harper & Row, 1960.

———: Bone smashing by late Miocene hominidae. *Nature, 218*:528, 1968.

Lederberg, J.: Exobiology: approaches to life beyond the earth. *Science, 132*:393, 1960.

Leigh, E. G.: The ecological role of Volterra's equations. In Gerstenhaber, M. (Ed.): *Some Mathematical Problems in Biology.* American Mathematical Society, Providence, Rhode Island, 1969.

Lewis, G. N., and Randall, M.: *Thermodynamics,* 2nd ed., revised by K. S. Pitzer and L. Brewer. New York, McGraw-Hill, 1961.

Lewontin, R. C. (Ed.): *Population Biology and Evolution,* a symposium, June 1967. Syracuse, New York, Syracuse University Press, 1968.

Lingenfelter, P. E.: Production of carbon 14 by cosmic-ray neutrons. *Rev Geophys, 1*:35, 1963.

Longsworth, L. G.: Diffusion in liquids. In Gray, D. E. (Ed.): *American Institute of Physics Handbook,* 2nd ed. New York, McGraw-Hill, 1957, sect. 2, p. 205.

Lorenz, K. Z.: *King Solomon's Ring.* New York, Thomas Y. Crowell, 1952, p. 181.

————: *On Aggression,* trans. by M. K. Wilson. New York, Harcourt, Brace & World, 1966.

Lotka, A. J.: *Elements of Physical Biology,* reprinted in 1956 as *Elements of Mathematical Biology.* New York, Dover, 1924a, pp. 218-225.

————: *Elements of Physical Biology,* reprinted in 1956 as *Elements of Mathematical Biology.* New York, Dover, 1924b, Ch. VII.

————: *Elements of Physical Biology,* reprinted in 1956 as *Elements of Mathematical Biology.* New York, Dover, 1924c, Ch. VI.

MacArthur, J. W., and Baillie, W. H. T.: Metabolic activity and duration of life. *J Exp Zool, 53*:221 and 243, 1929.

MacBride, R.: *The Automated State.* Philadelphia, Chilton, 1967.

Malthus, T. R.: *Three Essays on Population.* New York, The New American Library, 1824 (reprinted in 1960).

Mantoux, P.: *The Industrial Revolution in the Eighteenth Century.* New York, Harper & Row, 1961, Ch. 3.

Marcson, S. (Ed.): *Automation, Alienation, and Anomie,* a collection of contributions. New York, Harper & Row, 1970.

Margullis, L.: Symbiosis and evolution. *Sci Am, 225*:49, 1971.

Markham, C. (Ed.): *Jobs, Men, and Machines: Problems of Automation,* a conference in New York, May 15, 1963. New York, Frederick A. Praeger, 1964.

Mason, B.: *Principles of Geochemistry.* New York, John Wiley and Sons, 1952.

Mathews, M. V.: *The Technology of Computer Music.* Cambridge, M. I. T. Press, 1969.

May, R. M.: Limit cycles in predator-prey communities. *Science, 177*:900, 1971.

McCay, C. M.: Chemical aspects and the effects of diet upon ageing. In Lansing, A. I. (Ed.): *Cowdry's Problems of Ageing.* Baltimore, Williams and Wilkins, 1942, p. 680.

Meadows, D. L., Meadows, D. H., Behrens, W. W., and Randers, J.: *The Limits to Growth: A Report for the Club of Rome's Project on the Predicament of Mankind.* New York, Universe Books, 1972.

Mendel, G.: *Mendel's Principles of Heredity,* trans. by W. Bateson in 1909. Cambridge, Cambridge University Press, 1865.

Minorsky, N.: *Nonlinear Oscillations.* Princeton, New Jersey, D. Van Nostrand, 1962, Ch. 3.

Morowitz, H. J.: *Energy Flow in Biology.* New York, Academic Press, 1968, p. 2.

Morse, P. M., and Feshbach, H.: *Methods of Theoretical Physics.* New York, McGraw-Hill, 1953, vol. II, p. 1584.

Motz, L., and Duveen, A.: *Essentials of Astronomy.* Belmont, California, Wadsworth, 1966a, p. 605.

———: *Essentials of Astronomy.* Belmont, California, Wadsworth, 1966b, p. 608.

Muller, H. J.: Induced mutations in *Drosophila. Cold Spr Harbor Symp Quant Biol, 40*:151, 1941.

Neyman, J., Park, T., and Scott, E. L.: Struggle for existence, the *Tribolium* model: biological and statistical aspects. In Neyman, J. (Ed.): *Proceedings of the Third Berkeley Symposium of Mathematical Statistics and Probability.* Berkeley, University of California Press, 1956, vol. 4, p. 41.

Nicolis, G., and Babloyantz, A.: Fluctuations in open systems. *J Chem Phys, 51*:2632, 1969.

Nilsson, N. J.: *Learning Machines.* New York, McGraw-Hill, 1965:

———: *Problem-Solving Methods in Artificial Intelligence.* New York, McGraw-Hill, 1971.

Odum, E. P.: *Fundamentals of Ecology,* 2nd ed. Philadelphia, W. B. Saunders, 1959, p. 56.

Odum, H. T. *Environment, Power, and Society.* New York, Wiley-Interscience, 1971.

Oliver, J. A.: *The Natural History of North American Amphibians and Reptiles.* Princeton, D. Van Nostrand, 1955.

Olsson, I. U. (Ed.): *Radiocarbon Variations and Absolute Chronology,* Proceedings of the 12th Nobel symposium at Uppsala, Sweden, August 1969. Stockholm, Almquist and Wiksell, 1971.

Onsager, L.: Reciprocal relations in irreversible processes. *Phys Rev, 37*: 405, 1931.

Oparin, A. I.: Coacervate drops as models of prebiological systems. In Kimball, A. P., and Oro, J. (Eds.): *Prebiotic and Biochemical Evolution.* Amsterdam, North-Holland, 1971, p. 1.

Osterle, J. F.: A unified treatment of the thermodynamics of steady-state energy conversion. *Appl Sci Res, 12*:425, 1963.

Pardee, A. B., and Ingraham, L. L.: Free energy and entropy in metabolism. In Greenberg, D. M. (Ed.): *Metabolic Pathways.* New York, Academic Press, 1960, vol. 1, pp. 24-39.

Patten, B. C. (Ed.): *Systems Analysis and Simulation in Ecology*. New York, Academic Press, 1972, vol. II.

Pearl, R.: *The Rate of Living*. New York, Knopf, 1928.

———, and Reed, L. J.: On the rate of growth of the population of the United States since 1790 and its mathematical representation. *Proc Natl Acad Sci, 6*:275, 1920.

———: The biology of death—VII. natural death, public health, and the population problem. *Sci Mon, 13*:193, 1921.

Pielou, E. C.: *An Introduction to Mathematical Ecology*. New York, Wiley-Interscience, 1969.

Polunin, N. (Ed.): *The Environmental Future*, proceedings of the international conference on environmental future in Finland, summer of 1971. New York, Harper & Row, 1971.

Post, R. F.: Prospects for fusion power. *Phys Today, 26*:31, 1973.

Prigogine, I.: *Introduction to Thermodynamics of Irreversible Processes*, 2nd ed. New York, John Wiley & Sons, 1961.

———: Structure, dissipation and life. In Marois, M. (Ed.): *Theoretical Physics and Biology*. Amsterdam, North-Holland, 1969, p. 23.

Putnam, P. C.: *Power from the Wind*. New York, D. Van Nostrand, 1948.

Rashevsky, N.: *Looking at History Through Math*. Cambridge, M. I. T. Press, 1968.

Rayleigh, Lord of, J. W. Strutt: *Theory of Sound*. New York, Dover, 1894 (reprinted in 1945), vol. 1, p. 102.

Reeder, W. G., and Cowles, R. B.: Aspects of thermoregulation in bats. *J Mammal, 32*:389, 1951.

Reynolds, J.: *Windmills and Watermills*. New York, Praeger, 1970.

Ribbands, C. R.: *The Behavior and Social Life of Honeybees*. London, Bee Research Association Limited, 1953, Ch. 34.

Richardson, L. F.: *Statistics of Deadly Quarrels*. Pacific Grove, California, Boxwood Press, 1960a, p. 176.

———: *Statistics of Deadly Quarrels*. Pacific Grove, California, Boxwood Press, 1960b, p. 291.

———: *Statistics of Deadly Quarrels*. Pacific Grove, California, Boxwood Press, 1960c, p. 290.

———: *Statistics of Deadly Quarrels*. Pacific Grove, California, Boxwood Press, 1960d, p. 148.

Rosenzweig, M. L.: Paradox of enrichment: destabilization of exploitation ecosystems in ecological time. *Science, 171*:385, 1971.

Rubner, N.: *Mitt Geschich Med Wein*, 9(7):58, 1908.

Sagan, C.: On the origin and planetary distribution of life. *Radiat Res, 15*: 174, 1961.

Sage, B. H.: *Thermodynamics of Multicomponent Systems*. New York, Reinhold, 1965, p. 63.

Sansone, G., and Conti, R.: *Non-Linear Differential Equations*, revised ed., trans. by A. H. Diamond. New York, Macmillan, 1964.

Saunders, J. F.: A cryobiologist's conjeeture of planetary life. *Cryobiology,* 6:151, 1969.

Sawyer, J. S.: Man-made carbon dioxide and the "greenhouse" effect. *Nature, 239:23,* 1972.

Sayre, K. M., and Crosson, F. J.: *The Modeling of Mind, Computers and Intelligence.* Notre Dame, Indiana, University of Notre Dame, 1963.

Schroedinger, E.: *What is Life and Other Essays.* New York, Doubleday, 1956, pp. 67-84.

Schwartz, A. W.: In Busby, D. E. (Ed.): *Recent Advances in Aerospace Medicine.* Dordrecht, Holland, D. Reidel, 1971, p. 48.

Scott, E. L., and Bolz, R. W. (Eds.): *Automation and Society,* a symposium. The Center for the Study of Automation and Society, Athens, Georgia; vol. 1, 1969; vol. 2, 1970.

Scott, J. W.: Mating behavior of the sage grouse. *Auk, 59:*477, 1942.

Sears, F. W.: *An Introduction to Thermodynamics, the Kinetic Theory of Gases, and Statistical Mechanics,* 2nd ed. Cambridge, Addison-Wesley, 1953, p. 200.

Shneour, E. A., and Ottesen, E. A.: *Extraterrestrial Life: An Anthology and Bibliography.* Publication 1296A, National Academy of Science—National Research Council, Washington, D.C., 1966.

Siders, R. A.: *Computer Graphics; A Revolution in Design.* New York, American Management Association, 1966.

Silberman, C. E.: *The Myths of Automation.* New York, Harper & Row, 1966.

Slobodkin, L.: *Growth and Regulation of Animal Populations.* New York, Holt, Rinehart, and Winston, 1963.

Small, G. L.: *The Blue Whale.* New York, Columbia University Press, 1971. *See* especially Chapter entitled "Economics and Extermination."

Smith, F. E.: Population dynamics in *Daphnia magna* and a new model for population growth. *Ecology, 44:*651, 1963.

Soodak, H. (Ed.): *Reactor Handbook.* New York, John Wiley and Sons, 1962, vol. III, part A.

Statistical Office of the United Nations: *Statistical Yearbook of the United Nations, 1969.* New York, United Nations, 1970, p. 40, Table 9.

Stommel, H.: *The Gulf Stream.* Berkeley, University of California Press, 1965.

Strehler, B. L.: *Time, Cells, and Aging.* New York, Academic Press, 1962.

Stryer, L.: Optical asymmetry. In Pittendrigh, C. S., Vishniac, W., and Pearman, J. P. T. (Eds.): *Biology and Exploration of Mars.* Publication 1296, National Academy of Science—National Research Council, Washington, D.C., 1966, pp. 141-146.

Stuart, E. B., Gal-Or, B., and Brainard, A. J. (Eds.): *A Critical Review of Thermodynamics.* Baltimore, Mono Book Corporation, 1970.

Sverdrup, H. U., Johnson, W., and Fleming, R. H.: *The Oceans.* New York, Prentice-Hall, 1942, p. 675.

Terborgh, G.: *The Automation Hysteria.* New York, W. W. Norton, 1966.

Tinbergen, N.: *Social Behavior in Animals with Special Reference to Vertebrates.* London, Methuen, 1966.

U.S. Bureau of the Census: *Historical Statistics of the United States, Colonial Times to 1957.* U.S. Bureau of the Census, Washington, D.C., 1960.

Usher, A. P.: *A History of Mechanical Inventions.* Boston, Beacon Press, 1929a (reprinted in 1959), p. 342.

———: *A History of Mechanical Inventions.* Boston, Beacon, 1929b (reprinted in 1959), p. 352.

Utida, S.: Cyclic fluctuations of population density intrinsic to the host-parasite system. *Ecology, 38*:442, 1957.

Van Lawick-Goodall, J.: *In the Shadow of Man.* Boston, Houghton-Mifflin, 1971.

Verhulst, P. F.: Recherches mathématique sur la loi d'accroissement de la population. *Mem Acad R Med Belg,* Ser. 2, *18*:1, 1844.

———: Deuxième mémoire loi d'accroissement de la population. *Mem Acad R Med Belg,* Ser. 2, *20*:1, 1846.

Volterra, V.: *Mem Acad Lincei, 2*:1, 1926.

———: *RC Acad Lincei, 5*:463, 1927.

———: *Leçons sur la Théorie mathématique de la Lutte pour la Vie.* Paris, Gauthier-Villars, 1931.

Walker, G.: *J Mech Eng Sci, 4*:266, 1962.

Washburn, S. L.: Speculations on the interrelations of the history of tools and biological evolution. In White, L. A. (Ed.): *The Evolution of Man's Capacity for Culture.* Detroit, Wayne University Press, 1959, p. 21.

Watt, K. E. F.: The effect of population density on fecundity in insects. *Can Entomol, 92*:674, 1960a.

———: *Can Entomol, 92*:129, 1960b.

Weiner, B.: *Theories of Motivation from Mechanism to Cognition.* Chicago, Markham, 1972.

Welte, D.: Balance and cycle of carbon. In Wedepohl, K. H. (Ed.): *Handbook of Geochemistry,* Berlin, Springer-Verlag, 1969, vol. II/1, pp. 6-O-2.

Wesley, J. P.: Background radiation as the cause of fatal congenital malformation. *Int J Radiat Biol, 2*:97, 1960.

———: Frequency of wars and geographical opportunity. *J Conflict Resolution, 6*:387, 1962.

———: *The Application of Some Thermodynamic Principles to Estimate the Likelihood of Life in the Solar System, Part I.* A report to NASA (NGR 26 004 014), 1966.

———: Thermodynamic estimate of the likelihood of life in the solar system. *Curr Mod Biol, 1*:214, 1967.

———: Frequency of wars and geographical opportunity. In Pruitt, D. G., and Snyder, R. C. (Eds.): *Theory and Research on the Causes of War.* Englewood Cliffs, New Jersey, Prentice-Hall, 1969, p. 229.

Wilson, A. H.: *Thermodynamics and Statistical Mechanics.* Cambridge, Cambridge University Press, 1957.

Wilson, C. L., and Matthews, W. H. (Eds.): *Man's Impact on the Global Environment: Assessment and Recommendations for Action.* Cambridge, M.I.T. Press, 1970.

Woolhouse, H. W. (Ed.): *Evolution.* Number VII, a symposium at Oxford, July 1952. New York, Academic Press, 1953.

Wynne-Edwards, V. C.: *Animal Dispersion in Relation to Social Behavior.* New York, Hafner, 1962.

———: Self-regulating systems in populations of animals. *Science, 147:* 1543, 1965.

Zaikin, A., and Zhabotinsky, A.: Concentration wave propagation in two-dimensional liquid-phase self-oscillating system. *Nature, 225:535,* 1970.

# INDEX